Handbook of Technical Textiles

Handbook of Technical Textiles

Editor

Michael Kunek

scitus
academics

Handbook of Technical Textiles
Edited by **Michael Kunek**

ISBN: 978-1-68117-672-7
Library of Congress Control Number: 2016946538

SCITUS Academics LLC,
616, Corporate Way, Suite 2, 4766,
Valley Cottage, NY 10989

www.scitusacademics.com

Notice

Printed in United States of America on Acid Free Paper

Preface

Technical Textiles are welfare textiles used for their diverse and multifunctional properties. Technical textiles offer several advantages in their functional aspects for improving health and safety, cost effectiveness, and durability and strength of textile material. These performance-enhancement products are applicable in the protective clothing, agriculture, medical, infrastructure development, automotive, aerospace, sports, and packaging sectors. It is a large and growing sector and supports a vast array of other industries. The global growth rates of technical textiles are about 4% per year greater than the growth of home and apparel textiles, which are growing at a rate of 1% per year. Currently, technical textile materials are most widely used in filter clothing, furniture, hygiene medicals and construction material. An exceptional feature of technical textiles is the use of innumerable varieties of raw materials, processes, products and applications for their production. Technical textiles are reported to be the fastest growing sector of the textile industrial sector. At the present time, space and aerospace textiles are making a significant contribution to the increasing market for textiles.

Handbook of Technical Textiles provides readers with a comprehensive understanding of the latest advancements in technical textiles. It provides widespread coverage of the manufacture, processing and applications of high tech textiles for a huge range of applications. Technical textiles have never been a single coherent industry sector and market segment. It is developing in many different directions with varying speeds and levels of success.

Table of Contents

Chapter 1. Smart Technical Textiles Based on Fiber Optic Sensors 1

Chapter 2. The use of laser in garment manufacturing: An overview 37

Chapter 3. Medical Smart Textiles Based on Fiber Optic Technology: An Overview 65

Chapter 4. Medical Textiles as Vascular Implants and Their Success to Mimic Natural Arteries 91

Chapter 5. Assessment of Firefighters' needs for personal protective equipment 129

Chapter 6. Systems thinking in designing automotive textiles 149

Chapter 7. Wearable Wireless Cardiovascular Monitoring Using Textile-Based Nanosensor and Nanomaterial Systems 173

Chapter 8. RFID in textile and clothing manufacturing: technology and challenges 197

Chapter 9. Influence of Immersion Conditions on The Tensile Strength of Recycled Kevlar®/Polyester/Low-Melting-Point Polyester Nonwoven Geotextiles through Applying Statistical Analyses 225

Chapter 10. **Introducing Textiles as Material of Construction of Ethanol Bioreactors** **253**

Chapter 11. **Development of plasma enhanced antiviral surgical gown for healthcare workers** **271**

Index **289**

CHAPTER 1

Smart Technical Textiles Based on Fiber Optic Sensors

Katerina Krebber[1]

[1] Federal Institute for Materials Research and Testing (BAM) Berlin, Germany

1. INTRODUCTION

Smart technical textiles are by definition textiles that can interact with their environment. They can sense and react to environmental conditions and external stimuli from mechanical, thermal, chemical or other sources. Such textiles are multifunctional or even "intelligent" which is fulfilled by a number of sensors incorporated in the textiles. The embedded sensors are sensitive to various parameters such as temperature, strain, chemical, biological and other substances.

Technical textiles are commonly used within several industrial sectors ranging from medical, healthcare, earthworks, construction, civil engineering, transport, to name a few. Europe has driven substantial developments in technical textile technologies[1]. Smart technical textiles are going to stimulate the European engineering, transportation and construction industry and to improve human performance and health. For example, technical textiles are extensively used in construction in form of geotextiles for the reinforcement of earthworks and masonry structures. The retrofitting of existing masonry walls and soils structures by technical textiles gains more and more importance especially in connection with earthquake protection of historic buildings and protection of roads and railway embankments against landslides. Wearable health systems and protective clothing have been recognized as key technologies to improve the personal protection and health care of Europe's citizens[2]. Smart biomedical garments and clothing act as "a second skin" and detect, for instance, vital signals of the wearer's body or changes in the wearer's environment.

The most effort in the past was made to integrate non-optical sensors into textiles. Optical fibers integrated in textiles were mostly explored for illumination or luminescent purposes. Smart technical textiles containing fiber optic sensors are still an exception. When integration of sensors into textiles is considered, optical fibers have a serious advantage over other kinds of sensors due to their fibrous nature. The optical fiber is similar to textile fibers and can be ideally processed like standard textile yarns. Particularly, the integration of polymer optical fibers (POF), with their outstanding material properties, into technical textiles has not seriously been considered, until now. POF offer additional benefits to users. They are lightweight, robust, cheap and easy to handle. Especially because of their high elasticity and high breakdown strain POF are ideally suited for integration into technical textiles[3].

2. GEOTEXTILES BASED ON DISTRIBUTED FIBER OPTIC SENSORS FOR STRUCTURAL HEALTH MONITORING

For stabilization and reinforcement of geotechnical structures like dikes, dams, railways, embankments, landfills and slopes geotextiles are commonly used. The incorporation of optical fibers in geotextiles leads to additional functionalities of the textiles, e.g. monitoring of mechanical deformation, strain, temperature, humidity, pore pressure, detection of chemicals, measurement of the structural integrity and the health of the geotechnical structure (structural health monitoring). Especially solutions for distributed measurement of mechanical deformations over extended areas of some hundred meters up to some kilometers are urgently needed. Textile-integrated distributed fiber optic sensors can provide for any position of extended geotechnical structures information about critical soil displacement or slope slides via distributed strain measurement along the fiber with a high spatial resolution of less than 1 m. So an early detection of failures and damages in geotechnical structures of high risk potential can be ensured.

Geotextiles with incorporated fiber optic sensors based on fiber Bragg gratins (FBG) were demonstrated in the past[4]. Monitoring systems based on such geotextiles can only measure quasi-distributed strain over limited lengths and the relative high price of the FBG-equipped geotextiles might be an additional drawback of the systems. The monitoring of extended geotechnical structures like dikes, dams, railways, embankments or slopes requires sensor technologies with gauge lengths of some hundred meters or even more. Sensor systems based on the stimulated Brillouin scattering in silica fibers have been used for such monitoring purposes. It was reported in the past about a geotextile-based monitoring system using the Brillouin optical-fiber frequency-domain analysis (BOFDA) for measurements of critical soil displacements of dikes[5]. However, the excellent measurement technique based on Brillouin scattering in silica fibers reaches its limits when strong mechanical deformations, i.e. strain of more than 1 % occurs. In such a case sensors based on silica fibers cannot be reliably used. Furthermore, silica fibers are very fragile when installing on construction sites and, therefore, special robust and expensive glass fiber cables have to be used. For that reason, the integration of POF as a sensor into geotextiles has become very attractive because of the high elasticity, high breakdown strain and the capability of POF of measuring strain of more than 40 %. Especially the monitoring of relative small areas with an expected high mechanical deformation such as endangered slopes takes advantage of the outstanding mechanical properties of POF. The monitoring of slopes is a very important task in the geotechnical engineering for prevention of landslide disasters and no reliable sensor methods exist, so far. To overcome the limit of glass-fiber-based geotextiles, a novel distributed fiber optic sensor based on low-priced standard POF and using the OTDR (optical time-domain reflectometry) which is suitable for integration in technical textiles has been developed and demonstrated[6].

Such innovative textile-integrated distributed Brillouin and POF OTDR sensors for the above mentioned monitoring purposes have been developed within several German projects and the European project POLYTECT. The POLYTECT project has particularly focused on the development of polyfunctional technical textiles against natural hazards. The aim of POLYTECT has been to develop and investigate new multifunctional textile structures for the application in construction for the retrofitting of masonry structures and earthworks. The retrofitting of existing masonry walls and soil structures is particularly important for earthquake protection of historic buildings and protection of earthworks against landslides. The new and advanced textile structures containing optical fibers as sensors will be able to increase the ductility and the structural strength of masonry and geotechnical structures and to prevent structural damage[6]. For this, the sensors incorporated into the textile structures will monitor strain, deformation, humidity and will detect presence of chemicals. The development of the sensors carried out within the above mentioned projects has advanced and a number of field tests using distributed Brillouin and POF OTDR sensors have successfully been conducted.

2.1. Monitoring Of Geotechnical Structures Using Distributed Brillouin Sensors Embedded In Geotextiles

The use of stimulated Brillouin scattering (SBS) for distributed measurement of temperature and strain was already demonstrated 20 years ago[7]. The SBS is the most dominant nonlinear effect in single-mode silica fibers and can be described as a three-wave-interaction of two contra-propagating light waves and an acoustic wave in the fiber. Because of the strain and temperature dependence of the Brillouin frequency shift of the scattered light, sensor systems based on this effect can be used for distributed strain and temperature measurements. The first distributed Brillouin sensing systems named Brillouin optical-fiber time-domain analysis (BOTDA) operated in a time-domain, which means that a short pulse is sent along the fiber and the backscattered light is recorded over time and contains information about the strain or temperature along the fiber[8]. During the last two decades the performance of BOTDA sensor systems has improved steadily. The operating range of these sensors is typically in the order of 20-30 km for 2-3 m spatial resolution. Today, several devises based on this technique are commercially available.

In 1996 an alternative approach named Brillouin optical-fiber frequency-domain analysis (BOFDA) was introduced[9]. The BOFDA operates with sinusoidally amplitude-modulated light and is based on the measurement of a baseband transfer function in frequency domain by a network analyzer (NWA). A signal processor calculates the inverse fast Fourier transform (IFFT) of the baseband transfer function. In a linear system this IFFT is a good approximation of the pulse response of the sensor and resembles the strain and temperature distribution along the fiber (Fig. 1). The frequency-domain method offers some advantages compared to the BOTDA concept. One important aspect is the possibility of a narrow-bandwidth operation in the case of BOFDA. In a BOTDA system broadband measurements are necessary to record very short pulses, but in a BOFDA system the baseband transfer function is determined point-wise for each modulation frequency, so only one frequency component has to be measured by NWA with a narrow resolution bandwidth. The use of a narrow bandwidth operation (detectors) improves the signal-to-noise ratio and the dynamic range compared to those of a BOTDA sensor without increasing the measurement time. Another important advantage of a BOFDA sensor is that no fast sampling and data acquisition techniques are used. This reduces costs. Particularly, the low-cost-potential of BOFDA sensors is very attractive for industrial applications.

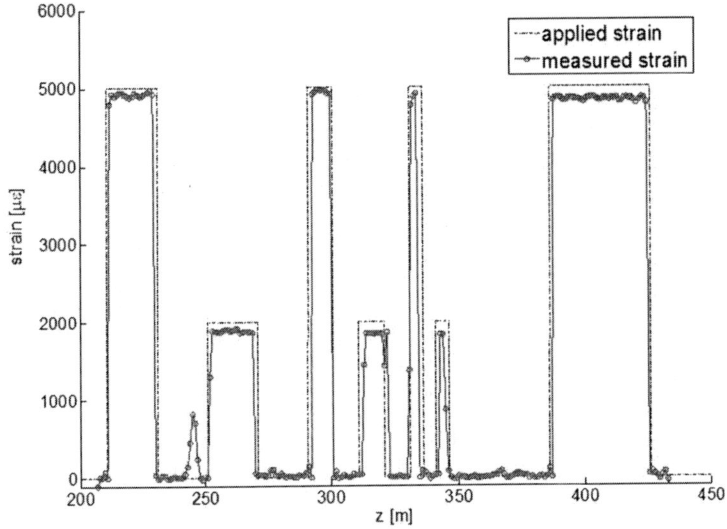

Figure 1. Distributed strain profile measured on a single-mode silica fiber using BOFDA.

As already pointed out, distributed Brillouin sensors are well qualified for the distributed monitoring of mechanical deformation (strain) of extended geotechnical structures like dikes, dams and highways of lengths of some hundred meters up to some kilometers and no alternative sensor techniques for such monitoring purposes exist so far. To push the development of such sensor systems in connection with innovative monitoring solutions based on smart technical textiles, several research projects have been running in Germany and Europe. The German research program RIMAX (Risk Management of Extreme Flood Events) has mainly focused on the development of intelligent monitoring systems for dike protection and was launched as a consequence of extreme floods in Germany in the past decade. A low-cost monitoring system based on the BOFDA technique and geotextiles containing silica fibers as distributed Brillouin sensors have been developed within the program[10].

Geotextiles are commonly used in dikes for reinforcement of the dike body and erosion prevention. By embedding sensing optical fibers in the textiles, distributed measurements of critical mechanical deformations/soil displacements of dikes of several kilometers can be realized. So an early detection of failures in dikes, dams and other large geotechnical structures can be ensured in order to prevent a total collapse of these structures in case of natural disasters. An important task when considering integration of optical fibers in geotextiles is to ensure an accurate transfer of the mechanical quantities to be measured, i.e. of strain, from the soil to the textile and so to the fiber. For this, a stable and damage-free integration of the optical fibers in the geomats is of essential importance. The Saxon Textile Research Institute (STFI) e.V., Chemnitz, Germany has developed a technology to integrate optical fibers into geotextiles so that the sensing fiber is well affixed onto the textile and the integration procedure does not affect the optical and sensing properties of the fibers. Also the use of special coating and cable materials are of crucial importance to protect the fragile single-mode silica fibers against fiber-breakage during the integration into the textiles and the installation on construction sites. For that, a novel glass fiber cable was developed and manufactured by Fiberware, Mittweida, Germany to fulfill the above mentioned requirements on robustness and to assure accurate strain transfer to the sensing fibers[11]. Fig. 2 shows the special cable as well as different types of geotextiles with embedded glass fiber cables.

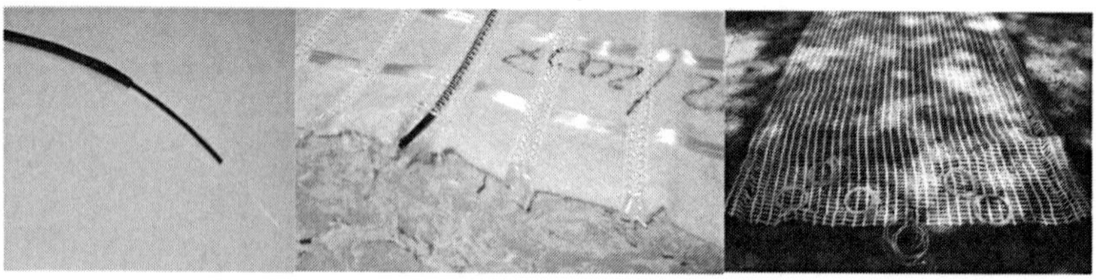

Figure 2. Special glass fiber cable for strain sensing manufactured by Fiberware, Germany (left) and two different types of geotextiles (middle: nonwoven geotextile, right: geogrid) manufactured by STFI, Germany with embedded glass fiber cables.

The BOFDA monitoring system has been optimized to fit the demands on dike monitoring: detection of mechanical deformation (strain) with a spatial resolution of 5 m over a distance range of up to 10 km. The functionality of the monitoring system and the fiber-sensors-equipped geotextiles has been proven in several installations and field tests in dikes and dams. For example, Fig. 3 shows the installation of geotextiles with embedded Brillouin sensing fibers in a gravity dam in Solina, Poland. A thin soil layer of several 10 cm put onto the geomats after installation has been proven to be a sufficient protection of the textile-integrated glass fiber cables against heavy machinery and construction work.

An application-like test was carried out at a laboratory dike (15 m long) at the University Hannover, Germany[11]. A sensor-based geotextile was installed on top of the dike and was covered with a thin soil layer. To simulate a mechanical deformation/soil displacement, a lifting bag was embedded into the soil and was inflated by air pressure. This induced a break of the inner slope of the dike and a soil displacement (Fig. 4). The soil displacement was clearly detected and localized by the BOFDA system.Fig. 5 shows the distribution of the mechanical deformation (strain) in the dike measured by the BOFDA system at two different air pressure values.

Figure 3. Installation of a non-woven geotextile containing single-mode silica fibers as Brilloin sensors in a gravity dam in Solina, Poland.

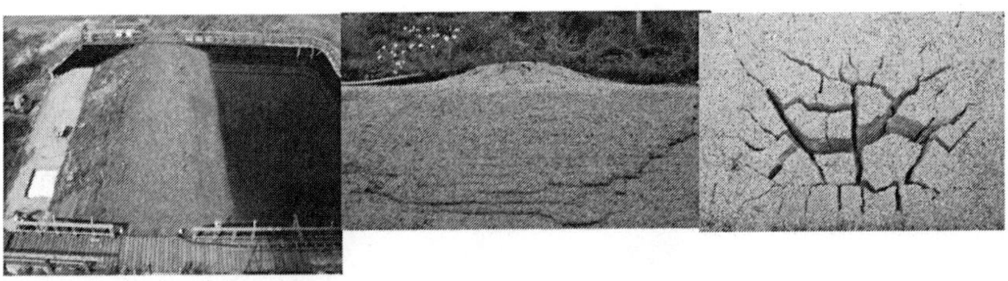

Figure 4. Laboratory dike at the University Hannover, Germany and soil displacement in the dike.

As previously mentioned, a geotextile with embedded Brillouin sensing fibers was installed in a gravity dam in Solina, Poland to prove the feasibility of the whole concept in the framework of a real field test Fig. 6. The goal of the field test was to detect possible geophysical activities in the dam by the fiber-sensor-equipped geotextile of a length of 17.5 m manufactured by STFI, Germany and embedded in the soil. Distributed measurements by using a commercially available BOTDA system from Omnisens were conducted. Fig. 7 shows the distributed Brillouin frequency shift measured on the fiber section embedded in the soil. In the fiber sections between 205 m and 240 m (where the geomat was embedded in the soil) a mechanical load is assumed which results in a change of the recorded Brillouin frequency in these fiber sections.

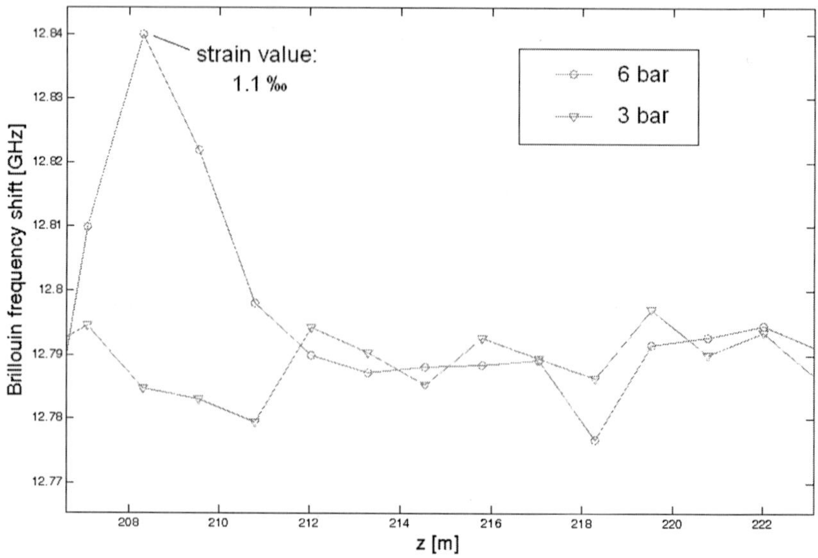

Figure 5. Detection of a soil displacement (strain) in the laboratory dike shown in Fig. 4 using the BOFDA system.

Figure 6. Gravity dam in Solina, Poland (left) and the construction site with the sensor-based non-woven geotextile before embedding in the soil and 3 years later (right).

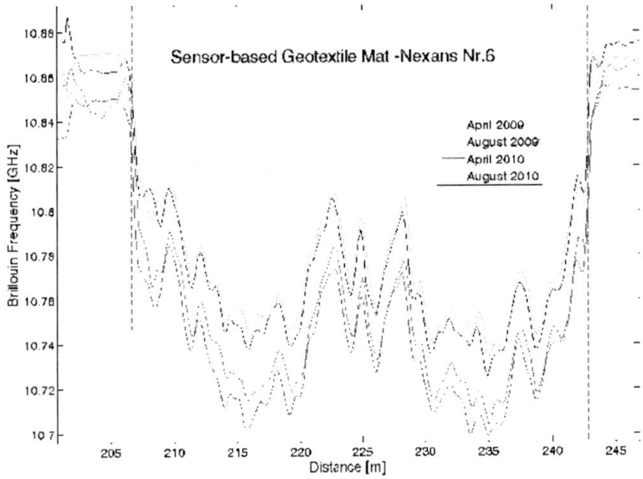

Figure 7. Distributed Brillouin frequency shift measured on the fiber section embedded in the soil (between 205 m and 240 m) 3 years after installation of the geomat.

With the objective of a cost-effective optimization of the BOFDA system a novel measurement concept based on a digital signal processing has been realized[11]. This concept employs a novel digital data acquisition technique, which takes advantage of the reduced bandwidth required in BOFDA sensor systems. The backscattered optical signals can be digitally sampled using state-of-the-art analog-to-digital converters and is processed off-line by means of modern digital signal processing methods, avoiding complex and expensive analog components such as filters, oscillators and circuitry for signal analysis. The digital optical signal processing features several advantages compared to the measurement process using NWA: less hardware is required, an increase of the dynamic range due to the offline signal processing and improvement of the data acquisition time is expected.

2.2. Monitoring Of Geotechnical Structures Using Distributed Pof Otdr Sensors Embedded In Geotextiles

To overcome the limit of silica-fiber-based distributed sensors, a novel distributed strain sensor based on low-priced standard POF and using the

OTDR technique for monitoring of mechanical deformations of geotechnical structures has been developed. Already published results showed that it is possible to measure distributed strain in POF using the OTDR technique[12]. In the framework of the German research project "Sensitive textile structures" (within the German program "ZUTECH" – "Future technologies") and the European project POLYTECT further investigations of this effect with respect to the development of a new, distributed POF sensor embedded in technical textiles have been performed[6], [13].

The functional principle of the POF OTDR technique is very simple. An optical pulse is launched into the fiber and the backscattered light mainly caused by Rayleigh scattering is recorded as a function of time. The time interval from launching the pulse into the fiber until the return of the backscattered light (pulse response) depends linearly on the distance of the scattering location. The level of the backscattered light increases at locations where strain is applied to the POF. Fig. 8 (left) shows the OTDR response of an unstretched POF (solid line) and of a stretched POF (broken line) which is stretched at about 42 m on a 1.4 m long section by 16 %. Fig. 8 (right) shows the relative change of scattering of the stretched POF section at different strain values between 0 % and 16 % (calculated relative to the scattering of the unstretched fiber). The scattered light increases steadily with applied strain. Today, several OTDR devices for POF are commercially available on the market. In the described investigations a photon counting OTDR device from Sunrise Luciol has been used. The device operates at 650 nm, has a dynamic range of 35 dB and allows a measurement of Rayleigh scattering along a length of more than 100 m. The photon counting technique is ideal for achieving high dynamic range on very short sensing lengths. The two-point spatial resolution of the OTDR device is limited to 10 cm. An additional solution to evaluate the strain or length change of a fiber section is to evaluate the shift of reflection peaks along the fiber (see Fig. 8, left). Such peaks originate for example from Fresnel reflections at the fiber end or fiber connectors. This technique provides an absolute length change measurement with a resolution of up to 1.5 mm.

Fig. 9 shows the increase of the scattered light versus applied strain. A non-linear dependence between the OTDR signal (the backscattering) and the applied strain in the whole strain range was obtained. Strain of up to 45 % was measured using standard PMMA POF.

The attenuation of standard PMMA POF limits the distance range of distributed POF OTDR sensors to about 100 m. Low-loss perfluorinated POF show a big potential as distributed strain sensors for long distances[14]. It has been shown that using perfluorinated POF it is possible to monitor fiber lengths of more than 500 m (Fig. 10, left). Recent research has demonstrated, that perfluorinated POF allow the measurement of very high strain values of up to 100 % (Fig. 10, right).

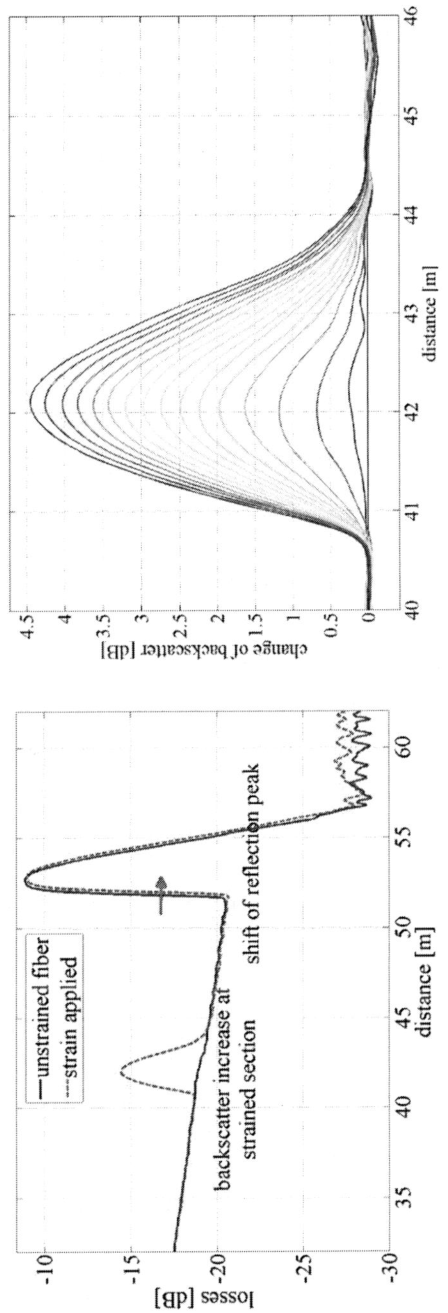

Figure 8. Left: OTDR trace of POF in unstretched condition (solid line) and of POF with a stretched fiber section at about 42 m (broken line). Right: Change of the scattering along a 1.4 m long POF section that is stretched from 0 to 16 % in steps of 1 %.

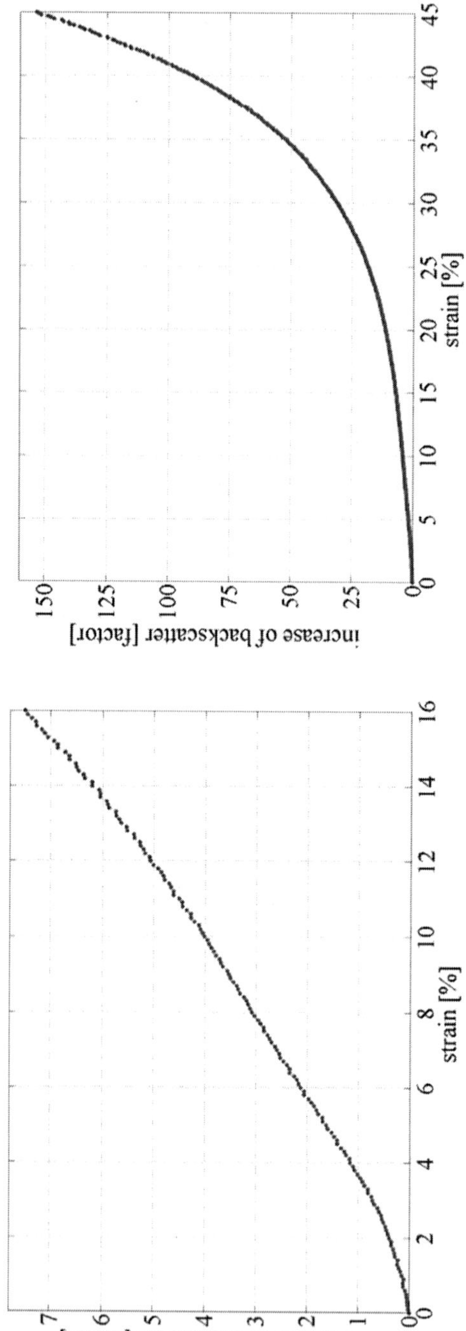

Figure 9. Change of the backscattering as a function of strain measured on standard PMMA POF.

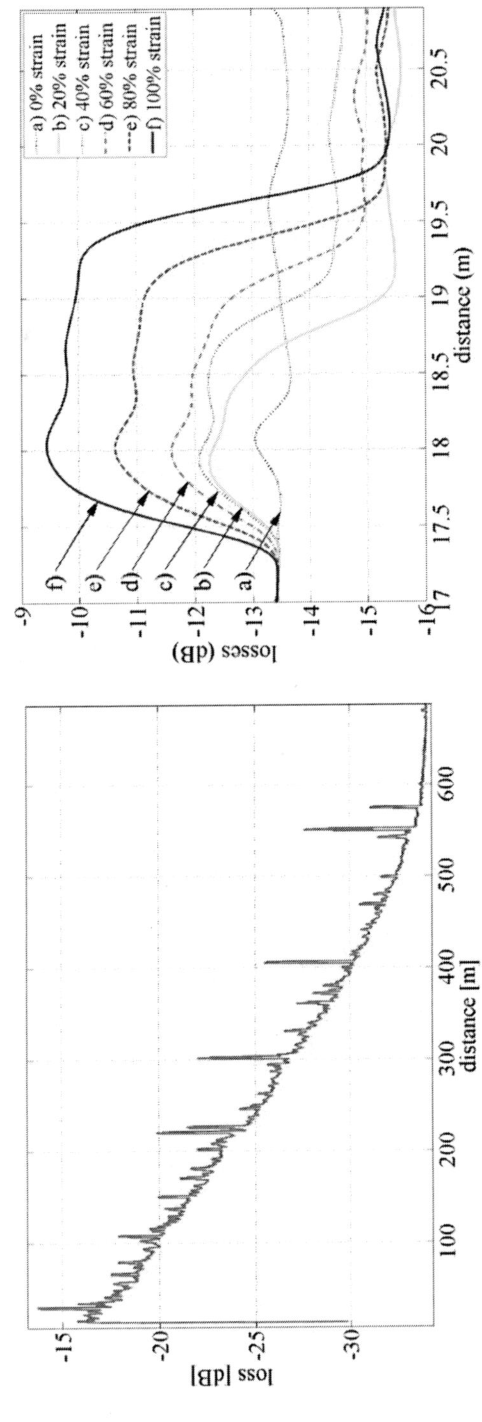

Figure 10. Left: OTDR trace of perfluorinated POF. Right: OTDR signal of perfluorinated POF strained up to 100 %.

Technologies for a damage-free integration of POF into different types of geotextiles have successfully been developed and demonstrated by several textile partners in Europe like STFI, Germany and Alpe Adria Textil, Italy. Fig. 11 shows the integration of POF into nonwoven geotextiles at STFI e.V. as well as a geogrid containing POF. Already the first field tests proved that the POF-equipped geotextiles are suited for installation on construction sites. POF-based geomats have successfully been installed in a railway embankment near Chemnitz, Germany (Fig. 12)[6]. All POF sensors have survived the installation on construction site without any damage. Their functionality has been regularly tested (Fig. 12, right).

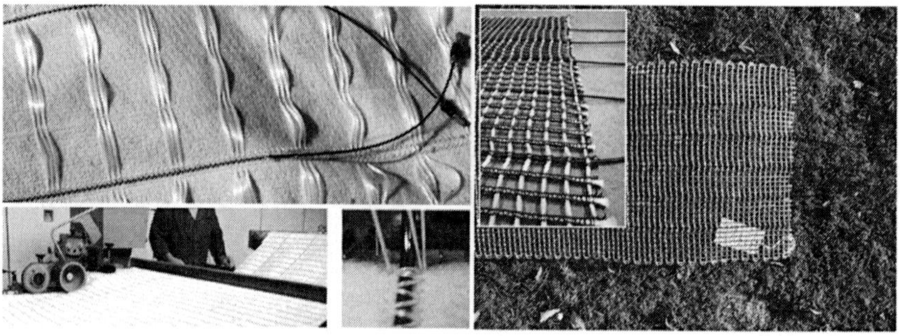

Figure 11. Integration of POF into nonwoven geotextiles at STFI e.V. (left) and a geogrid containing POF (right).

During the last years, the POF-equipped geotextiles have successfully moved from the laboratory to the field. Several field tests have successfully been conducted, e.g. in an open brown coal pit near Belchatow, Poland[15]. The test was initiated, organized and supervised by Gloetzl Baumesstechnik GmbH, Germany in close cooperation with Budokop, Poland and the owner of the coal pit. A sensor-equipped geogrid was installed directly on top of a creeping slope. The 10 m long geogrid was manufactured by Alpe Adria Textil, Italy and comprised one standard PMMA POF. Fig. 13 shows the installation of the sensor textile on top of the slope. It is covered with a 10 cm thick sand layer. The textile is installed with the POF sensor bridging the cleft perpendicular to the opening. The geogrid was installed in a slightly corrugated way simulating realistic installation conditions.

Figure 12. Installation of POF-equipped geotextiles in a railway embankment near Chemnitz, Germany (left and middle) and OTDR traces measured on the textile-integrated POF (right).

Figure 13. Installation of a geogrid containing PMMA POF at a creeping slope in a brown coal pit near Belchatow, Poland.

Measurements were conducted before and after installation. Fig. 14 (left) shows the OTDR traces of the sensor fiber section (the magnitude of the backscatter increase relative to a reference measurement) in the middle of the textile where the fiber bridges the cleft. The figure clearly shows a backscatter increase due to strain in the fiber at the position where the cleft was opening. The high peak at about 35 m is caused by a very high and confined strain in the sensor fiber and textile. The magnitude of the backscatter increase corresponds to a maximum strain in the fiber of more than 10 %. Such high strain values can only be measured by POF sensors. Silica fiber-based sensor systems would have failed at a strain exceeding about 1 %.

Due to the gradual increase of cleft width, the overlying textile and therefore the sensor fiber change their absolute length. By evaluating the relative shift of the reflection peaks at both ends of the textile-integrated fiber, the values of the total elongation of the fiber sensor indicating the width of the cleft was obtained. Fig. 14 (right) shows a relative linear increase of the POF length with time. The measurements indicate that the creep velocity of the slope was constant during the time of observation with an average rate of about 2 mm per day.

Recently, novel geogrids containing low-loss perfluorinated POF (PF POF) have been developed and manufactured by Alpe Adria Textil. Already the first field test has proved that the PF POF-equipped geotextiles are suited for installation on construction sites. PF POF-based geomats have successfully been installed at the creeping slope Kap Arkona at the German Baltic coast (Fig. 15, left). All PF POF sensors have survived the installation on construction site without any damage. At present, their functionality has been regularly tested by using the POF OTDR technique (Fig. 15, right).

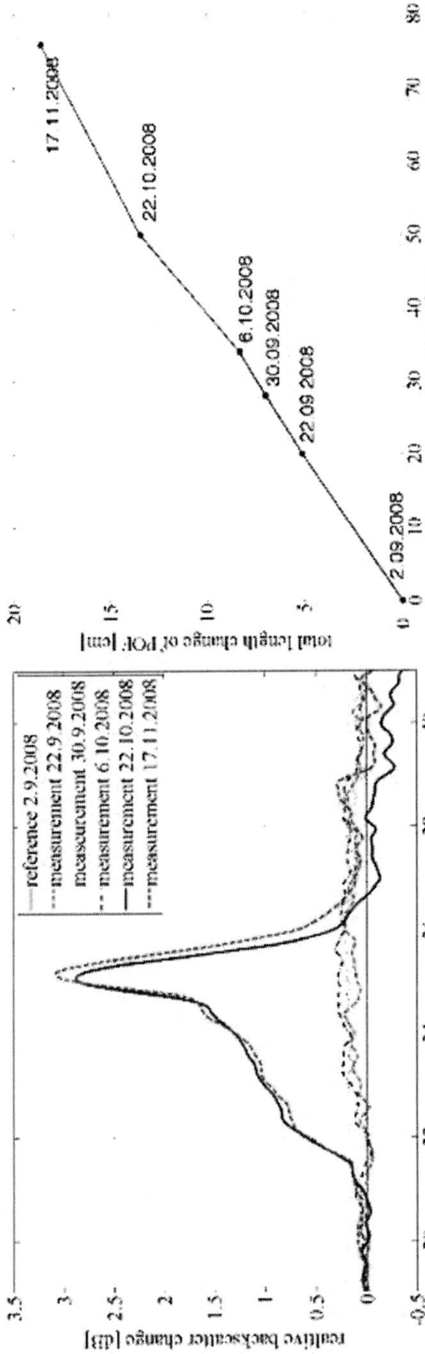

Figure 14. POF OTDR traces at the position of the cleft (left) and total elongation of the POF obtained by a peak-shift evaluation (right).

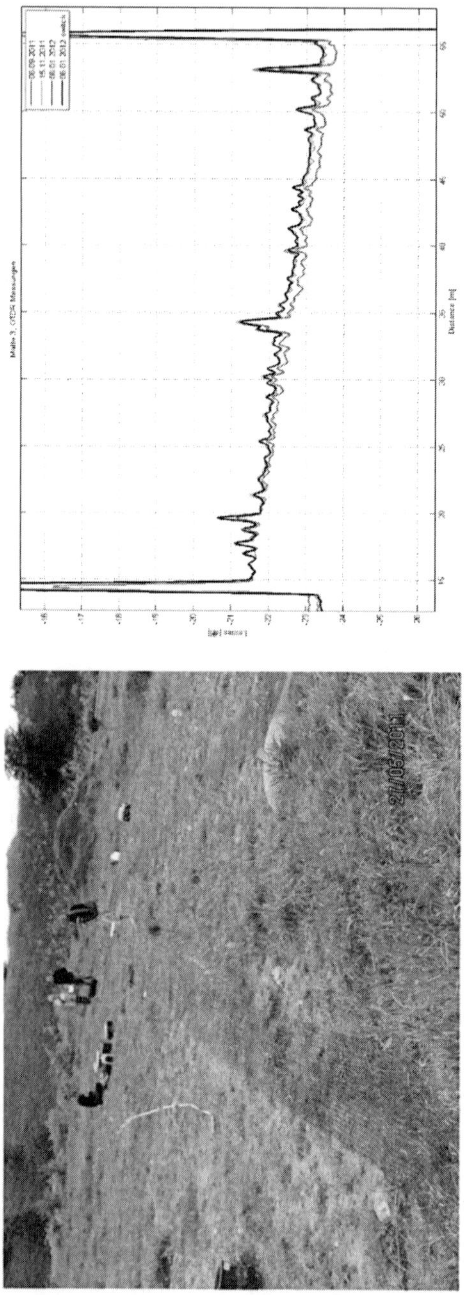

Figure 15. Left: Installation of PF POF-equipped geotextiles at the creeping slope Kap Arkona at the German Baltic coast. Right: OTDR traces measured on the textile-integrated PF POF after installation.

The successful demonstration of the distributed POF OTDR sensors in the field and the huge interest of the geotechnical industry in these sensors resulted in the development of the first commercially available product based on distributed POF sensors – GEDISE: Distributed Sensor Technique in Geotextiles using POF (Fig. 16). GEDISE is commercially available by Glötzl GmbH, Germany.

2.3. Monitoring Of Masonry Structures Using Distributed Pof Otdr Sensors Embedded In Technical Textiles

The motivation to monitor masonry structures by sensor-equipped technical textiles is to strengthen the masonry body and enhance the ductility of the structures and at the same time to monitor the structural health and detect any damage of the structures, e.g. due to earthquakes. The development of sensor-based technical textiles containing fiber optic sensors for the retrofitting of masonry structures is an innovative task of the European project POLYTECT. The targeted applications are masonry and heritage structures that are structurally vulnerable, for example in earthquake regions. Typical structural damages that have to be detected are vertical cracks. POF sensors are very promising for that since they not only enable distributed strain measurement, they are also appropriate to detect very short strained fiber sections of a few millimeters that will occur in case of cracks. For example, Fig. 17shows the monitoring of a crack opening in a masonry structure using a POF OTDR sensor. A technical textile containing POF was applied to the surface of the masonry sample[6]. Using the POF OTDR technique it was possible to detect a crack opening of 1 mm and also the increase of the crack width up to 20 mm in steps of 2 mm (Fig. 17, right).

Using the POF OTDR technique a field test was conducted on an one-storey brick building on a seismic shaking table[16]. The test was organized and supervised by the Institute of Mechanics of Materials and Geostructures (IMMG), Greece. Fig. 18 shows the POF sensors bonded to the wall with a cementitious resin matrix. The testing procedure included several strong shocks, which resulted in structural damage of the building. The task of the distributed POF sensors was to provide information about the existence and location of cracks in the structures. The occurred cracks were detected and localized with the POF OTDR sensor. Fig. 19 shows the OTDR traces measured on one sensor-fiber which was installed diagonally on the wall. Two cracks were detected by the sensor at the locations indicated in Fig. 18. The stronger signal at 27 m is caused by a 2 mm crack at the corner above the door. A smaller, almost invisible crack has been detected at 150 cm distance from the first crack at the lower right corner of the wall.

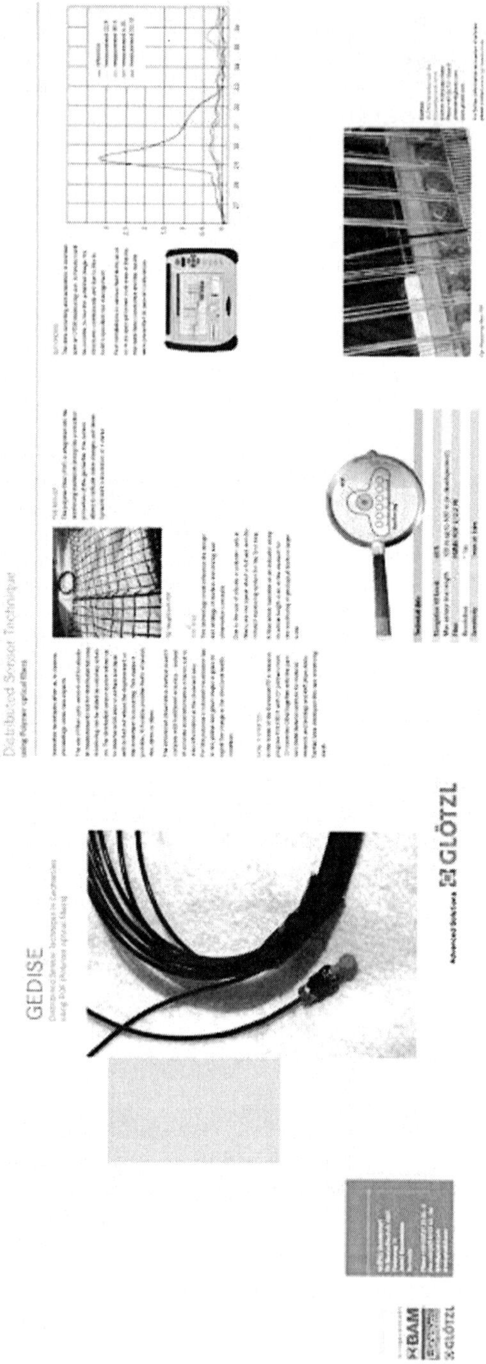

Figure 16. Leaflet of GEDISE: Distributed Sensor Technique in Geotextiles using POF (**www.gloetzl.de**).

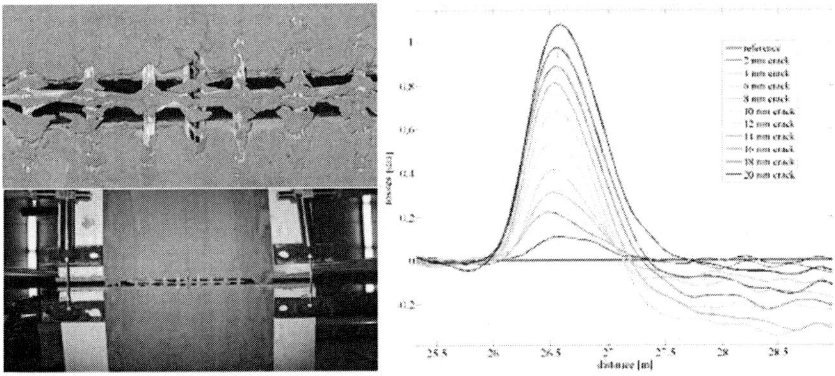

Figure 17. Monitoring of a crack opening in a masonry structure (Institute IfMB at the University of Karlsruhe, Germany) with a POF-equipped masonry textile (STFI e.V., Chemnitz, Germany). The right side of the figure shows POF OTDR backscatter signals at the location of the crack at different crack opening steps.

Figure 18. Brick building on a shaking table with POF sensors installed horizontally and diagonally.

During the last years, several field tests have successfully been conducted on real masonry buildings reinforced by POF-sensors-based technical textiles, one of them on a masonry house at the Eucentre in Pavia, Italy (Fig. 20). The testing procedures of the textile-equipped masonry building included several strong seismic shocks (simulating earthquakes) that resulted in several cracks in the masonry walls. The occurred cracks were clearly detected and localized by the distributed POF OTDR sensor (Fig. 21) which demonstrated the potential of this technique to be used also for damage detection of masonry and heritage structures.

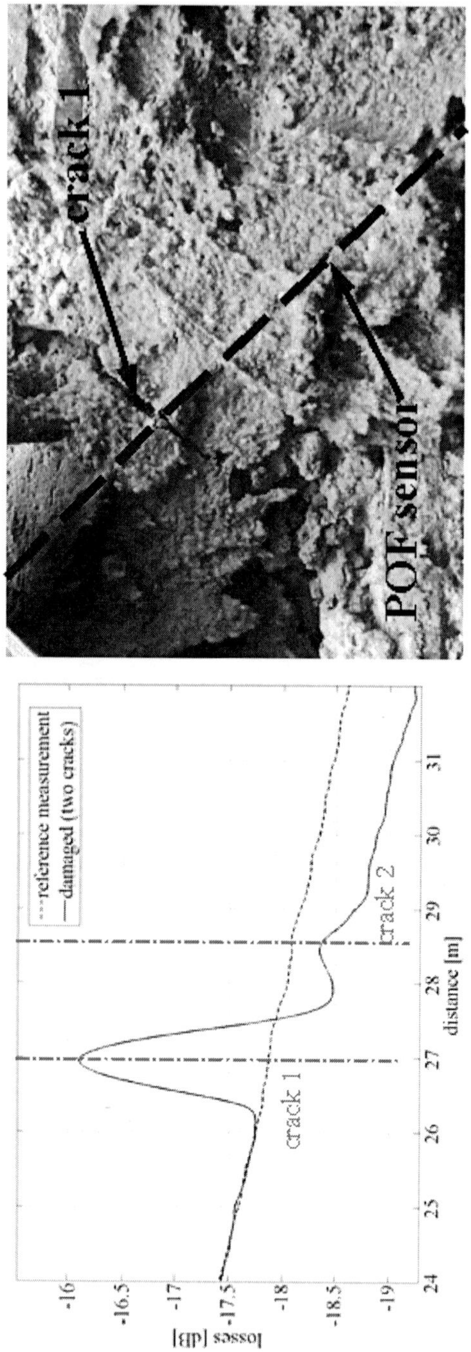

Figure 19. POF OTDR trace showing two cracks at 27.0 m and 28.5 m (left) and the corresponding first crack at 27.0 m of a width of 2 mm (right).

Figure 20. Application of technical textiles containing POF on a masonry building at the Eucentre in Pavia, Italy.

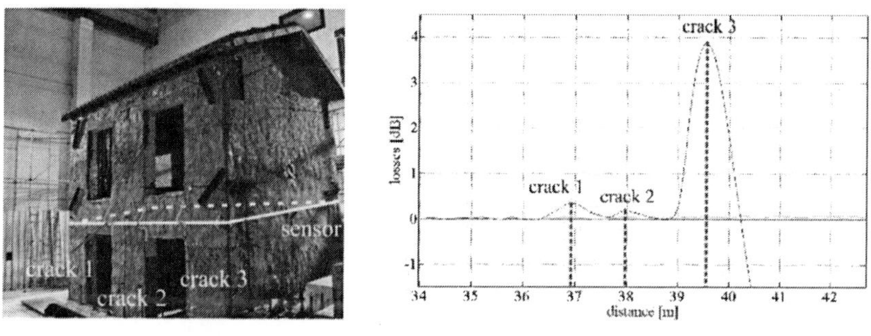

Figure 21. Detection of cracks in a masonry wall by a textile-embedded distributed POF OTDR sensor after several seismic shocks applied to the building.

3. MEDICAL TEXTILES BASED ON FIBER OPTIC SENSORS FOR HEALTHCARE MONITORING

Healthcare monitoring of patients and old people who require a continuous medical assistance and treatment is a subject of a number of research activities in Europe. In order to increase the mobility of such patients, the development of wearable monitoring systems able to measure important physiological parameters of the patients is targeted. Europe has considerably pushed the developments of such wearable biomedical clothing containing different types of sensors by a number of research projects.

The European project OFSETH (optical fiber sensors embedded into technical textile for healthcare) supported by the 6th European framework program, has investigated how various vital parameters such as respiratory movement, cardiac rate and pulse oxymetry can be measured by fiber optic sensors based on silica and polymer optical fibers, embedded into medical textiles. As a result, wearable solutions for healthcare monitoring, for patients requiring a continuous medical assistance and treatment, are available. Despite of already existing electrical and also fiber optic sensors, OFSETH has achieved a breakthrough in the healthcare monitoring by combining the advantages of pure fiber optic sensor technologies and wearability of the textiles and so increasing the functionality of the sensor and the comfort of the system.

The OFSETH developments have targeted in the first place on the monitoring of sedated or anaesthetized patients under Medical Resonance Imaging (MRI)[17]. In this case electrical sensors cannot play a role; fiber optic sensors are advantageous because of their electromagnetic compatibility. The use of fiber optic sensors instead of electrical sensors will reduce the electromagnetic disturbance of the MRI field. Additionally, metallic parts and conductive wires of electrical sensors cause burns on the patient's skin in the MRI field. Fiber optic sensors are free from such metallic components and so burning hazard for the patients can be prevented. Besides, fiber optic sensors offer the advantage that the monitoring unit can be placed out of the MRI field and can be connected to the sensor by a fiber cable of some five or ten meters.

Anaesthetized patients are usually transferred from the induction room to the MRI room and back under anesthesia. A continuous monitoring of the patients from the induction to the end of the anesthesia is required but in fact the medical staff usually uses different monitoring devices during the whole procedure, because the most standard monitoring devices are not transportable or not MRI compatible. After the MRI examination, the patient is transferred back, still anesthetized, in the worst case without any monitoring system which puts the patients at risk of anesthetic complications. Therefore, a transportable monitoring system, able to follow the patients from the induction room to the MRI room and back without being removed is needed. The wearability of such a system will increase its functionality and the comfort to the user. Wearable monitoring systems can also be used in the ambulatory healthcare monitoring and the monitoring of Sudden Infant Death Syndrome. Therefore, OFSETH has mainly addressed the textile integration issues and in this context has extended the capability of wearable solutions for healthcare monitoring.

For MRI applications there is especially need to monitor the patients' respiratory parameters: respiratory movement and respiratory rate. Therefore, OFSETH has focused, among other things, on the investigation of textile-integrated fiber optic sensors for respiratory monitoring of patients during MRI examinations. For this purpose, medical textiles that incorporate silica and polymer optical fibers have been investigated where a wearable, adaptable and MRI compatible monitoring system has been targeted.

The feasibility of using fiber optic sensors for respiratory monitoring was demonstrated in the past. It has been reported on fiber sensors woven into bandages or attached onto garments mainly using FBG (fiber Bragg gratings) and LPG (long period gratings) based on silica fibers[18], [19]. However, the poor compatibility of these sensors with industrial textile processes limits their flexibility and use for medical monitoring purposes.

Human breathing movement causes typical elongations of the abdominal circumference of adults of up to 3 %. Using silica fibers, limited strain values of up to 1 % can be measured. Therefore, with a special focus on using POF instead of silica fibers, OFSETH has investigated different fiber sensor techniques for respiratory monitoring[20], [21]. A highly important criterion for selecting POF as medical sensor is its biocompatibility, especially in case of fiber breakage.

3.1. Pof Otdr Sensor Embedded In Medical Textiles For Monitoring Of The Respiratory Movement

For the respiratory monitoring, there is an interest for the doctors to take information from both abdominal (for spontaneous ventilation) and thoracic (for intubated patients) movement[17]. Therefore, a distributed measurement of the respiratory signal, using only one monitor and one sensor fiber would be advantageous. Using an OTDR technique, it is possible to focus on a special part of the fiber and so to differentiate between abdominal and thoracic respiration. A distributed OTDR measurement makes possible to get only the required sensor information and to neglect loss contributions from non-sensing parts. In addition, an OTDR sensor system has the advantage of requiring only one fiber connection, which enables a quicker installation of the system on the patient.

A textile sample based on an elastic fabric containing a POF and manufactured by Centexbel and Elasta, Belgium was tested for the purposes of the respiratory movement monitoring by the OTDR technique. Since it was difficult to integrate a straight optical fiber into an elastic fabric, the textile sample uses a special macrobending sensor design developed by Multitel, Belgium (Fig. 22)[21]. The textile is divided in two sections: a short elastic part of about 10 cm whose length changes during the respiration and a longer non-elastic part. The POF is integrated into the elastic section to measure the elongation of the fabric due to the respiratory movement of the thorax or abdomen. The macrobending sensor design (described more detailed in Chapter 3.2) increases the sensitivity of the POF to the textile elongation and makes possible to detect small changes in the amplitude of the respiratory movement by the OTDR technique. Macrobending effects in POF induce changes of the backscattering in the corresponding area of the fiber that can be easily detected by the OTDR technique.

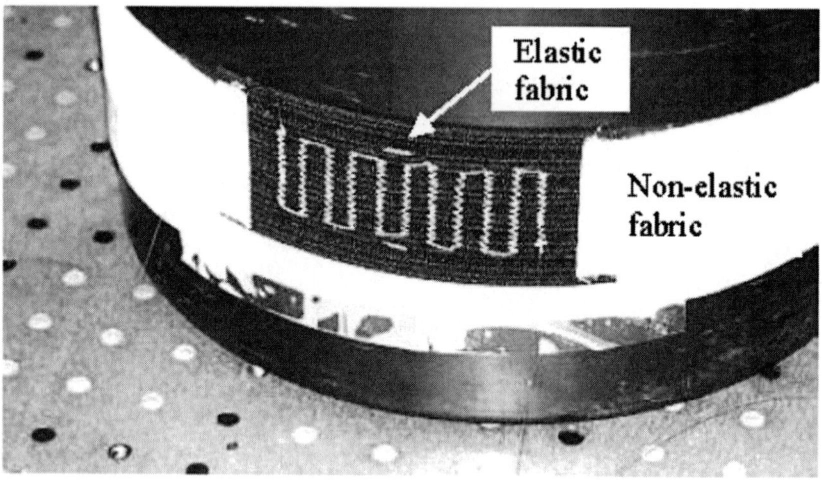

Figure 22. Textile sample containing a POF and based on the macrobending sensor design (textile: Centexbel & Elasta, Belgium; sensor design: Multitel, Belgium).

The feasibility of measuring the respiratory waveform and rate in real time by the POF OTDR technique was demonstrated on a healthy adult during normal breathing[21], [22]. The textile sample was attached around the abdomen of the adult and the elastic part of the textile was placed in the area experiencing the most elongation due to the breathing movement (Fig. 23). The sensor signal was acquired by a fast OTDR device produced by Tempo (OFM20), which operates at 650 nm wavelength, allows a two-point spatial resolution of 5 cm and has a dynamic range of > 20 dB. The device makes possible to measure an OTDR trace in less than 1 s with a sufficient SNR. This acquisition time is fast enough to measure normal human breathing. The changes of the abdominal circumference due to the breathing movement were recorded simultaneously. Fig. 23 shows the result and demonstrates the high potential of the POF OTDR technique for the considered monitoring purposes.

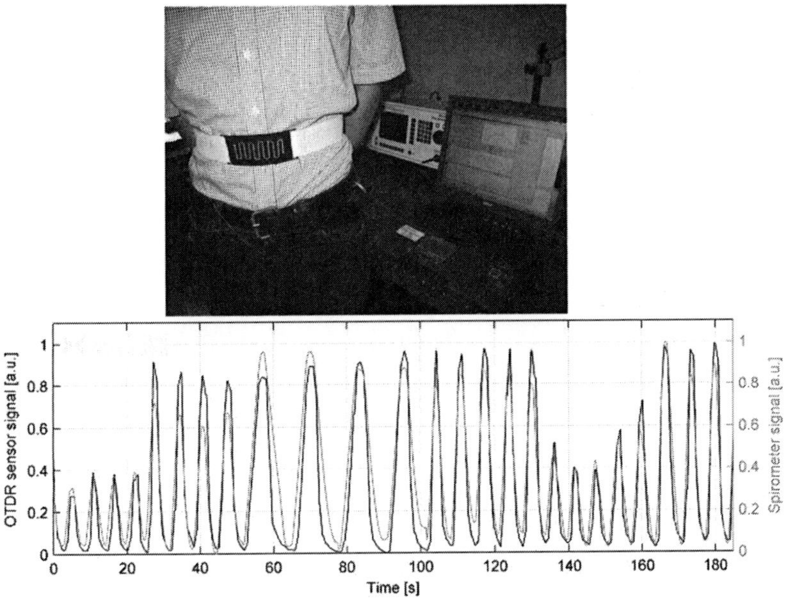

Figure 23. Monitoring of the respiratory abdominal movement of a human adult by POF OTDR sensor embedded in medical textiles (the OTDR sensor signal was compared with the signal measured by a spirometer).

3.2. Sensing Harness For Monitoring Of The Respiratory Movement

Considering the influence of different patient's morphology as well as textile integration issues to let free all vital organs for medical staff actions during incident or respiratory accidents, different fiber optic sensors have been integrated into a harness allowing an efficient handling and continuous measurement of the respiratory movement[22]. European norms in terms of textile and the medical specification have been taken into account for the design of the sensing harness where the fiber optic sensors are strategically placed for measurement of thoracic and abdominal movements caused by the breathing activity without corruption of one signal by another (Fig. 24). This design is composed of adjustable parts in order to fit the maximum of morphologies and to be worn both by men and women. The harness design keeps some places free, like the pre-cordium in order to facilitate resuscitation in case of cardiac arrest or hemodynamical failure, and give vital information on hemodynamical status during resuscitation. Access to the intra-venous infusion line has also been kept clear, for easy access during anesthesia or for resuscitation purpose. It has been ensured that there is no pressure on venous or arterial blood vessels which could obstruct the regular blood flow.

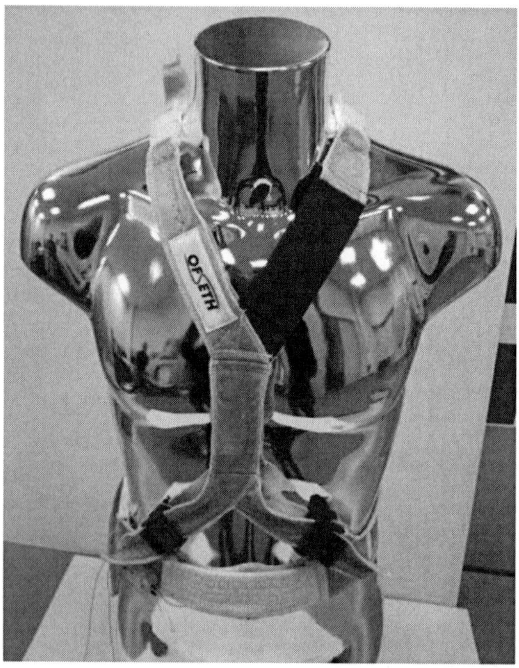

Figure 24. Sensing harness containing fiber optic sensors for the monitoring of patients under MRI. A thoracic respiration sensor is integrated in the black part (upper right); an abdominal respiration sensor is integrated in the white part (lower middle).

The elongation of the harness belt caused by the respiratory movement is measured using different fiber optic sensing principles based on FBGs and macrobending effects. The abdominal movement causes elongations of about 1-3%, which is much higher than for the thoracic movement which causes only a fractional percentage change. Therefore an FBG sensor which has high accuracy but a low strain limit is used for the thorax while for the abdomen a less accurate macrobending sensor is used which has a much higher strain limit.

The macrobending sensor developed by Multitel (Belgium) is based on bending effect of optical fibers (Fig. 25, left)[22]. Bends cause light coupling from guided modes into radiation modes and thus some power is lost. When the sensor textile is stretched, the curvature radius increases, and the bending loss decreases. Therefore the intensity variations at the output of the optical fiber will reflect the changes of the textile length, due to the respiratory movement. Macrobending sensors have the advantages that their interrogation is very simple: they require measurement of intensity changes, so the main components needed are an LED source and a photodiode. Standard single-mode silica fibers have been integrated into elastic fabrics, manufactured by Elasta (Belgium) during an industrial crochet fabrication process. The bending textile design has the advantage that the integration of the optical fibers into textiles is relatively simple. The bending design also ensures that the optical fiber is not damaged at high strain during integration. Due to the relatively high amplitude of the abdominal movement the signal-to-noise ratio is high enough to monitor the respiratory rate.

For the monitoring of the thoracic respiration movement an FBG sensor developed by Centexbel (Belgium) and Multitel is used. Due to the FBG inscription process the fiber sensor is weakened, which reduces facilities for integration of the fiber into the textiles. For this reason, only optical fibers with sufficient robustness should be used and conventional textile fabrication processes as opted for the macrobending sensor are inadequate for the FBG integration. The optical fiber containing the FBG was thus stitched directly onto an elastic fabric[22]. The robustness of the sensor is guaranteed by an additional silicone coating and polymer attachment points on both sides of the FBG are glued around the fiber for a better adhesion of the sensor onto the fabric and easy stitching without impairing the sensor properties (Fig. 25, right).

The harness based on the macrobending and FBG sensor was validated on a simulator in MRI environment[22]. A simulator based on a movable table was used (Fig. 26, left). The displacement of the table was realized by a balloon connected to the medical respirator allowing air-flow circulation by controlling the amplitude and frequency of the movement through the volume or air injected. The signals of the respirator, the fiber sensor response and the gradient signals emitted by the MRI were measured in real-time. Several configurations in terms of volume and/or frequency, in or out of the MRI tube and in presence of or without the MRI gradient were simulated and tested. As a result, it was demonstrated that the displacement of the movable table is detected in terms of amplitude and frequency. The signals of the fiber sensors were not degraded even when the system was submitted to the gradient of the MRI equipment in and out of the magnetic field (inside and outside the MRI tube respectively), as shown in Fig. 26 (right, large picture). At the same time, a clinical validation of the system was carried out at a hospital of Lille, France on several healthy volunteers and patients of the hospital's intensive care unit. Fig 26 (right, small picture) shows the typical signal patterns for both thoracic and abdominal movement detected by the textile-embedded FBG and macrobending sensor on healthy adults.

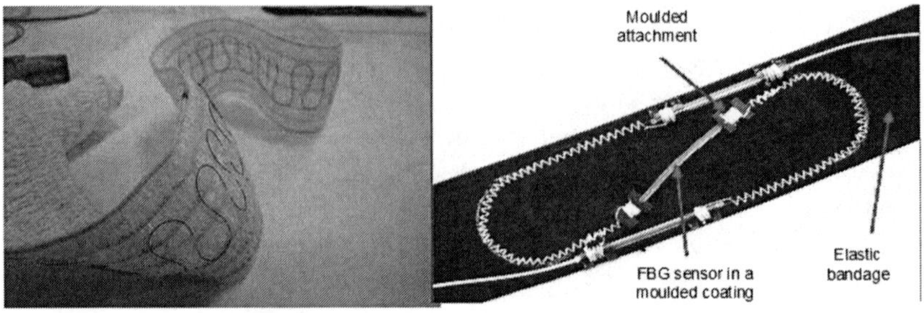

Figure 25. Left: Design of the macrobending sensor. A silica optical fiber was embedded into an elastic fabric during an industrial crochet fabrication process. Right: Design of the FBG sensor. A silica optical fiber containing an FBG was stitched onto an elastic textile.

3.3. Fiber Optic Sensors For Personal Protective Equipment

The European project i-Protect (intelligent PPE system for personnel in high-risk and complex environments) develops an advanced personal protective equipment (PPE) system that will ensure active protection and information support for personnel operating in high-risk and complex environments in firefighting, chemical and mining rescue operations[23]. The PPE system will be ergonomically designed and fully adapted to end-users' needs as well as to working conditions. The core of the project is the development of advanced materials and sensors to be used for a multi-functional PPE. This includes a real-time monitoring of risk factors (temperature, gas, oxygen level), users' health status (body temperature, respiratory rate, heart rate) and important protection parameters (end-of-service-life, air pressure in compressed units). The PPE will be wireless connected to a rescue command center.

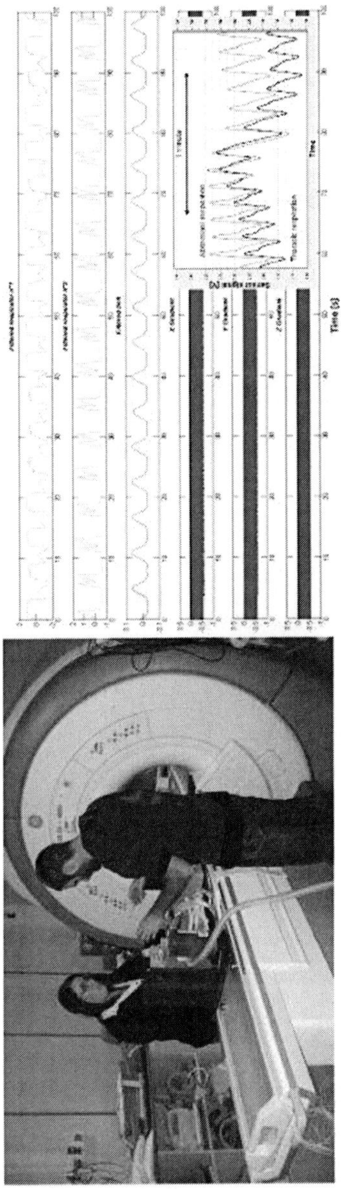

Figure 26. Left: Set-up of the MRI-compatible simulator of CIC-IT de Nancy, France. Right, large picture: Test of the FBG sensor in MRI environment. The first two curves are related to the respiration simulator, the third curve shows the FBG sensor response and the last three curves are related to the magnetic gradients of the MRI equipment. Right, small picture: Abdominal and thoracic respiration signals detected by the textile-embedded macrobending and FBG sensor during a clinical test on healthy volunteers.

For the monitoring of the users' health status smart underwear containing fiber optic sensors is being developed. Special attention is paid to the development of a heart rate sensor to be used as a textile-integrated sensor in underwear. A first sensor prototype is based on macrobending effects in POF[23]. The POF macrobending sensor is stitched onto an elastic fabric (the design is similar to this shown inFig. 25, left) and measures the small elongations of the textile which is caused by the heart movement. To increase the sensitivity of the sensor, the cladding of the POF was treated[23]. A sensor belt containing the POF sensor was tested on a healthy volunteer to measure the circumference changes due to the heart movement. The belt was wrapped around the chest of the volunteer close to the heart. Since the textile design is also sensitive to the respiratory movement, the POF macrobending sensor detects both the respiratory and heart rate at the same time (Fig. 27). A signal processing should be performed to filter the weak heart beats signals. It is expected that by using a modified textile and sensor design it must be possible to improve the sensitivity of the POF macrobending sensor to the heart movement. Alternatively, conventional monitoring techniques like plethysmography could be adapted for the purpose of such applications.

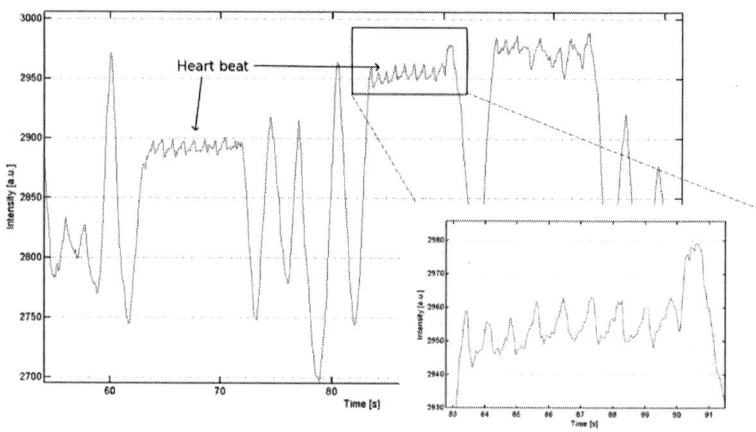

Figure 27. Monitoring of the heart rate of a healthy volunteer by using a textile-embedded POF macrobending sensor.

4. CONCLUSION

A number of research activities considering the development of novel smart technical textiles based on fiber optic sensors are running in Europe. Such smart technical textiles with embedded optical fibers are a potential new market niche for fiber optic sensors.

Several German projects and the European project POLYTECT have developed novel geotextiles with embedded distributed Brillouin and POF OTDR sensors for monitoring of geotechnical and masonry structures, providing an alarm signal in case of structural damage. Particularly sensors based on POF take advantage of the high robustness, high elasticity and high break-down strain of POF allowing distributed sensing of strong mechanical deformations of soil and masonry walls. Multifunctional, smart technical textiles incorporating fiber optics sensors are a cost-effective solution to increase the structural safety of such structures. The breakthroughs include the use of such textiles for reinforcement and at the same time for monitoring of earthworks and masonry walls, giving online information on the state and the performance of the structures and so preventing a total collapse. Such on-line and long-term monitoring systems will improve the chance of an early detection and the location of “weak points” and damages, and will make it possible to react rapidly and to control damages.

Novel monitoring systems based on medical textiles with embedded fiber optic sensors will be used at medium-term in the healthcare monitoring and for personal protection of rescues in high-risk environments where standard, non-optical monitoring systems show significant limits. Such medical textiles containing fiber optic sensors have been developed in the European projects OFSETH and i-Protect for the monitoring of the respiratory movement of anaesthetized patients under MRI and for the monitoring of the health status of rescues. Especially for MRI applications where transportable and MRI compatible devices are needed, pure fiber optic sensor solutions and the wearability of the textiles are advantageous. The design and comfort of such sensor systems will extend their use from hospitalization to the ambulatory healthcare monitoring and homecare.

NAN. ACKNOWLEDGEMENTS

The research has been carried out in the framework of the German projects "Sensorbasierte Geotextilien zur Deichertuechtigung” (FKZ 02WH0573) and “Sensitive Textilstrukturen” (AiF-Nr. 192 ZBG 1) as well as within the European projects POLYTECT (NMP2-CT-2006-026789), OFSETH (IST-

2004-027869) and i-Protect (NMP2-SE-2010-229275). The projects have received research funding from the Federal Ministry of Education and Research (BMBF), the Federal Ministry of Economics and Technology (BMWi) and the European Commission within the European 6th and 7thFramework Programs. The authors are thankful for the financial support and the fruitful cooperation with the project partners.

REFERENCES

1. “The future is Textiles”. Strategic Research Agenda of the European Technology Platform for the future of textiles and clothing, EURATEX (2006).

2. Lymberis, A. and A. Dittmar, A., “Advanced Wearable Health Systems and Applications”, IEEE Engineering in medicine and biology, pp. 29-33 (2007).

3. Krebber, K., Lenke, P., Liehr, S., Witt, J. and Schukar, M., “Smart technical textiles with integrated POF sensors”, Proc. SPIE 6933, 69330V-1 – 69330V-15 (2008).

4. Voet, M. R., Nances, A. and Vlekken, J., “Geodetect: a new step for the use of Fibre Bragg Grating technology in soil engineering”, Proc. of the 17th International Conference on Optical Fibre Sensors, 5855 (1), 214-217 (2005).

5. Noether, N., Wosniok, A., Krebber, K. and Thiele, E., “A distributed fiber optic sensor system for dike monitoring using Brillouin optical frequency domain analysis”, Proc. SPIE 6933, 69330T-1 – 69330T-9 (2008).

6. Liehr, S., Lenke, P., Wendt, M., Krebber, K., Seeger, M., Thiele, E., Metschies, H., Gebreselassie, B. and Muenich, J. C., “Polymer Optical Fibre Sensor for Distributed Strain Measurement and Application in Structural Health Monitoring”, IEEE Sensors J., 9 (11), 1330-1338 (2009).

7. Kurashima, T., Horiguchi, T. and Tateda, M., “Distributed-temperature sensing using stimulated Brillouin scattering in optical silica fibers”, Optics Letters 15 (18), 1038-1040 (1990).

8. Bao, X., Webb, D.J. and Jackson, D.A., "32-km distributed temperature sensor based on Brillouin loss in an optical fiber", Optics Letters 18 (18), 1561-1563 (1993).

9. Garus, D., Krebber, K., Schliep, F. and Gogolla, T., "Distributed sensing technique based on Brillouin optical-fiber frequency-domain analysis", Optics Letters 21(17), 1402-1404 (1996).

10. Noether, N., Wosniok, A., Krebber, K. and Thiele, E., "A Distributed fiber optic sensor system for dike monitoring using Brillouin frequency domain analysis," Proc. SPIE 7003, 700303 (2008).

11. Noether, N., Wosniok, A., Krebber, K. and Thiele, E., "A distributed fiber-optic sensor system for monitoring of large geotechnical strutures", Proc. of the 4th International Conference on Structural Health Monitoring on Intelligent Infrastructure (SHMII-4) (2009).

12. Husdi, I. R., Nakamura, K. and Ueha, S., "Sensing characteristics of plastic optical fibres measured by optical time-domain refelectometry", Meas. Sci. Technol. 15, 1553-1559 (2004).

13. Lenke, P., Liehr, S. and Krebber, K., „Improvement of the distributed strain sensor based on optical time domain reflectometry measurement in polymer optical fibers", Proc. of the 17th International Conference on Plastic Optical Fibre, (2008).

14. Liehr, S., Wendt, M., and Krebber, K., "Distributed strain measurement in perfluorinated polymer optical fibres using optical frequency domain reflectometry," Measurement Science and Technology 21(9), 094023–1 – 094023–6 (2010).

15. Liehr, S., Lenke, P., Wendt, M., Krebber, K., Gloetzl, R., Schneider-Gloetzl, J., Gabino, L. and Krywult, L., „Distributed Polymer Optical Fiber Sensors in Geotextiles for Monitoring of Earthwork Strutures", Proc. of the 4th International Conference on Structural Health Monitoring on Intelligent Infrastructure (SHMII-4) (2009).

16. Liehr, S., Wendt, M., Krebber, Muenich, J. C., Stempniewski, L. and Metschies, H., „Distributed polymer optical fiber sensors integrated in technical textiles for monitoring of masonry structures", Proc. of the 4th International Conference on Structural Health Monitoring on Intelligent Infrastructure (SHMII-4) (2009).

17. De jonckheere, J., Jeanne, M., Grillet, A., Weber, S., Chaud, P., Logier, R. and Weber, J. L., "OFSETH: Optical Fibre Embedded into technical Textile for Healthcare, an efficient way to monitor patient under

magnetic resonance imaging", Proc. IEEE EMBC conference Engineering in Medicine and Biology Society (2007).

18. Wehrle, G., Nohama, P., Kalinowski, H.J., Torres, P.I. and Valente, L.C.G., "A fibre optic Bragg grating strain sensor for monitoring ventilatory movements", Meas. Sci. Technol. 12, 805-809 (2001).

19. Allsop, T., Revees, R., Webb, D. J. and Bennion, I., "Respiratory monitoring using fibre long period grating sensors", Novel Optical Instrumentation for Biomedical Applications II, SPIE-OSA Biomedical Optics, SPIE 5864, Q1-Q6 (2005).

20. Grillet, A., Kinet, D., Witt, J., Schukar, M., Krebber, K., Pirotte, F. and Depré, A., "Optical Fiber Sensors Embedded into Medical Textiles for Healthcare Monitoring", IEEE Sensors J., 8 (7), 1215-1222 (2008).

21. Krebber, K., Grillet, A., Witt, J., Schukar, M., Kinet, D., Thiel, T., Pirotte, F. and Depré, A., "Optical fibre sensors embedded into technical textile for healthcare (OFSETH)", Proc. of the 16th International Conference on Plastic Optical Fibre, 227-233 (2007).

22. Witt, J., Narbonneau, F., Schukar, M., Krebber, K., De Jonckheere, J., Jeanne, M., Kinet, D., Paquet, B., Depré, A., D´Angelo, L.T., Thiel, T. and Logier, R., "Medical textiles with embedded fiber optic sensors for monitoring of respiratory movement", IEEE Sensors J., 12 (1), 246-254 (2012).

23. Witt, J., Krebber, K., Demuth, J. and Sasek, L., "Fiber optic heart rate sensor for integration into personal protective equipment", Proc. of International Workshop BioPhotonics 2011, Th6.26, 1-3 (2011).

CHAPTER 2

The use of laser in garment manufacturing: An overview

Rajkishore Nayak* and Rajiv Padhye

School of Fashion and Textiles, RMIT University,
25 Dawson St., Brunswick 3056, Australia

ABSTRACT

Laser is being used in apparel industry from nineteenth century for various garment manufacturing applications. There are several advantages of using laser over the conventional processes in cutting, engraving, embossing, denim fading and other applications. In addition, product damage potential is reduced, no/less consumables are needed and no problem of toxic by-product disposal as found in some processes. Today's laser equipment is a result of continuous research and development of earlier products, which has undergone several changes. The initial laser systems were cumbersome, hard to run and difficult to maintain. However, the modern laser systems are simpler in operation and maintenance. Furthermore, the earlier systems were involved with more safety issues and needed the gasses to be constantly replenished. The garment manufactures around the globe should take the advantage of laser application in the post multi-fibre agreement regime to make their products more competitive. This review focuses on the technology of laser including various classifications. In addition it includes the applications of laser in garment manufacturing, their potential hazards and health related concerns.

KEYWORDS

Laser Fabric cutting, Denim fading, Mass customization, Engraving

1. INTRODUCTION

Laser is being used in apparel industry from ninenteenth century. Recently the use of laser in apparel industry is increasing in cutting garment patterns, patterning designer neckties, 3D body scanning, denim fading and engraving leather (Nayak and Khandual 2010; Istook and Hwang 2001; Simmons and Istook 2003; Ortiz-Morales et al. 2003; Ozguney 2007; Bahtiyari 2011). The major reasons for wide application of laser in garment industries may be due to reduced cost, flexibility and anti-counterfeiting (Kovacs et al. 2006; Tarhan and Sarıışık, 2009; Yuan et al. 2012). For example the artwork of high-end necktie producers are digitally stored rather than physical patterns to lower the theft risk. When needed, the digital patterns are converted into physical samples using lasers (Lucas et al. 2015; Kan2015). Recently, the application of laser in denim engraving is increasing rapidly for value addition by replacing the traditional denim-distressing technics, which will take the denim segment to a height of sophistication that can never be realised by non-laser methods (Kan2014a). The unique nature of the garment manufacturing industry needs laser applications, which combines performance with reduced cost by eliminating the handling systems used in non-laser workstations.

Laser is an energy source, whose intensity and power can be precisely controlled. The laser beam can be focused to a desired object at specific angle depending on the application. Laser can help to cut a variety of objects ranging from flexible fabric to rigid and strong metal (Belli et al. 2005; Ondogan et al. 2005). Laser equipment are becoming widely popular in textile, leather and garment industries due to the advantage of accuracy, efficiency, simplicity and the scope of automation (Kan et al. 2010; Lu et al. 2010; Sutcliffe et al. 2000). For example the conventional cutting tools such as band blades, discs and reciprocating knives suffer from the limitations especially on delicate materials as the cutting force can displace the material, which can lead to inaccurate cutting (Nayak and Padhye 2015b). The traditional cutting methods often require an operator with full attention (Vilumsone-Nemes 2012). Hence, there is a trade-off between the maximum speed of cutting and the accuracy. In addition, other limitations include intricacy of the cut components, tool longevity and machine downtime during tool servicing. These limitations are not present in laser devices, which helps to achieve improved efficiency and reduced cost.

Laser cutting with processing speed, high precision, simple operation and other advantages, so in most industries can be used to, but it is in the clothing industry, leather processing plays a different role, can quickly cut leather graphics and draw precise clothing model (Potluri and Atkinson 2003; Ready et al. 2001). The benefit of laser cutting operations involve highly collimated beam that can be focused to a very fine dot of extremely high energy density for precise cutting. Garment industry pay attention to the size of the garment when processing precision, the purpose is to achieve high efficiency, exquisite tailoring, it is better than the traditional manual cutting by spectrum.

Although there are several applications of laser in apparel manufacturing, there are limited publications on this. Hence, an attempt was made in the current article to perform a thorough review that covers various applications and recent developments in laser that can provide guidelines for future directions. In addition, a systematic description has been done on the technology of laser including various types of laser. In addition, this review includes the applications of laser in garment manufacturing, their potential hazards and health related concerns.

2. LASER TECHNOLOGY

LASER is *light amplification by stimulated emission of radiation*. Laser is an electromagnetic radiation, produced by the atoms due to energy states are changed in some materials (Dowden 2009). The atoms promoted to higher energy states emit laser in the form of light by the process known as "stimulated emission". Subsequently, this laser is being amplified in a suitable lasing medium with the help of mirrors. The final laser delivers from the equipment as a stream similar to light. The colour of the laser depends on its wavelength. The most widely used unit used to express the wavelength of a laser is in nanometre (nm).

Laser light (Fig. 1) emitted from a laser has four fundamental characteristics: Intensity, Coherency, Monochromaticity, and Collimation, which distinguish it from natural light. A high energy concentration per unit area of the beam is present in the laser. A laser beam can be of very high intensity with 1–2 mm of beam diameter and an output power of some milliwatts (mW). All the lasers are not powerful in spite of the light with high intensity as intensity is estimated from the power output/area.

From Fig. 1 it can be observed that the ordinary light (from a thermal source) is incoherent, i.e., light waves are generated at different times and propagate in all possible directions randomly. However, laser is coherent as shown in the figure due to the waves are in phase while they propagate. Ordinary light is composed of all the colours in the visible region, but laser light is of a single colour or monochromatic. The coherent nature in addition to the monochromaticity results in highly collimated laser. As all the waves propagate in same phase in parallel lines, there is almost no divergence as observed in light. This property of laser helps in achieving high intensity even after travelling a long distance. The energy concentration of the beam can be increased by manifold when the beam is focused as a point with the help of optical lens.

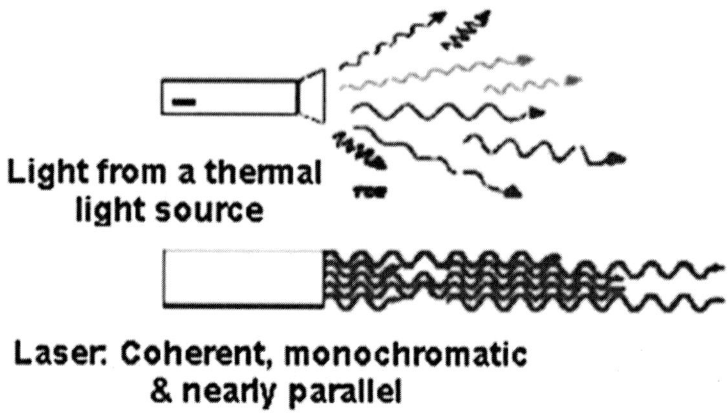

Figure 1. An ordinary light and a Laser beam

For the production of the laser beam a lasing material is stimulated with electric discharge in an enclosed container. A number of partial mirrors help in multiple internal reflections that lead to obtain the desired coherency and power output. Mirrors or other optical devices help in focusing the laser to a lens, which then reflects the laser to the working zone. In order to enable the cutting action by a laser away from the edge, it should be pierced before the laser cut using a high-power pulsed laser beam.

3. CLASSES OF LASING MEDIA

A substance, which can be excited by input energy, is called a lasing media. The lasing media have the ability to stay in a metastable state. It is generally transparent to light. The substance used as lasing media can be a solid, gas, liquid dye or semiconductor. Solid lasing media is used for producing solid-state laser. Due to the high density of lasing atoms, solid-state lasers can generate higher power output per unit volume compared to the gaseous-state lasers. A single gas or gaseous mixture can be used to produce gaseous-state lasers. Gaseous-state lasers are produced by passing the electric discharge in the gaseous lasing medium. The most common type of gaseous laser is produced from helium–neon (He–Ne). This laser is mainly used in teaching laboratories, at construction sites and supermarkets.

Liquid dyes can also produce laser by using ultraviolet (UV) light. These liquid dyes can be prepared by dissolving the solid dye particles in suitable solvent. The dyes and solvents with similar energy levels are selected, that help to form a continuum for achieving high energy output with wavelengths of visible light. Better tunability is the unique feature of a dye laser, which distinguishes it from the solid laser. A single dye produces a single wavelength whereas dyes can be mixed to obtain a range of wavelengths. Semiconductors can also be used to produce laser with the help of electric discharge. These lasers are characterised by their efficiency, miniature size and durability.

3.1 Laser classification based on hazards

Since the early 1970s, lasers were being classified into four classes and some subclasses depending on their wavelength and maximum output power. This classification was based on the severity of the damage to a person when exposed to laser. These classes can be from 1 to 4. A Class 1 laser is absolutely not dangerous when used, whereas the Class 4 laser is the most dangerous. The existing system was the revised system since 2002, prior to that the old system was used. In the new system amendments have been done to certain types of lasers having a lower hazard than mentioned in the old system. Based on the output power in a specific wavelength, lasers are classified. It is essential the correct information on the laser class, potential hazards and safety instructions are specified by the equipment manufacturers. In the classes 1–4 laser, there are the sub-classes 1M, 2M, 3A and 3B. The new system currently in use (shown in Table 1) uses Arabic numerals (1–4), whereas it was classified with Roman numerals (I–IV) in the old system.

Table 1. The new system for laser classification

Laser class	Features
Class 1	The Class 1 laser is the safest under normal use. These lasers may pose a risk when viewed with a telescope or microscope of sufficiently high aperture
Class 1M	The Class 1M laser is safe during normal use. However, when passed through a magnifying device such as a microscope can pose hazard. A laser falls in this class if the power that can pass through the pupil of a naked eye is lower than the accessible emission level (AEL) for Class 1
Class 2	A Class 2 laser is the visible-light laser (400–700 nm). It is safe as the blink reflex will limit the exposure time lower than 0.25 s. However, intentional holding of the blink reflex could lead to potential eye injury. Several measuring instruments and laser pointers are based on Class 2
Class 2M	This class laser is safe as the blink reflex if not viewed through optical instruments. Similar to Class 1M, this class laser lights are with a large divergence, the light passing through the pupil should not exceed the specifications for Class 2
Class 3A	This laser class is safe if handled carefully. The maximum permissible exposure (MPE) can be exceeded, which is associated with a low injury risk
Class 3B	This class is hazardous if exposed directly to the eye. Protective eyewear must be used where direct viewing is needed or may occur. The equipment with Class 3B lasers must be fitted with a safety interlock and a key switch
Class 4	This class is the most dangerous among all lasers. This laser can cause permanent eye damage or burn the skin as a result of direct, diffuse or indirect beam viewing or contact. These lasers may cause a fire risk as they can ignite combustible materials. Several laser used in scientific, industrial, military and medical applications fall in this category. The equipment with Class 4 lasers must be fitted with a safety interlock and a key switch

3.2 Classification based on generating media

Based on the generating media, lasers can be classified into three groups: (1) carbon dioxide (CO_2) lasers, (2) neodymium (Nd) lasers and (3) neodymium yttrium-aluminium-garnet (Nd-YAG) lasers (Thawari et al. 2005; Mathew et al. 1999; Schuocker 1989). The CO_2 lasers can be used for boring, cutting and engraving (Chow et al. 2012; Kan 2014b). CO_2 lasers can be of the four types: (a) fast axial flow, (b) slow axial flow, (c) transverse flow and (d) slab lasers (Powell 1993). In fast axial flow, a mixture of CO_2, nitrogen (N_2) and helium (He) is used at high speed by a turbine. A simple blower is used for slow axial flow and transverse flow lasers. Slab lasers need no pressurization as they work in static condition.

The Nd lasers and Nd-YAG lasers, although similar in style, they are applied in different areas (Schuocker 1989). The former is used for boring, whereas the later is used for engraving. All the laser classes can be used for welding.

Generally, CO_2 lasers are produced by passing the electric discharge through a gaseous mixture excited by direct current (DC) with radio frequency (RF) energy. This method is widely used as the DC-excited designs need electrodes inside the cavity, due to risk of erosion. Industrial cutting of steel, aluminium, paper, plastics, wood, and fabrics can be performed by CO_2

lasers. For cutting metals and ceramics; and scribing Nd-YAG lasers are used.

The laser generator and external optics need cooling while in use (Choudhury and Shirley2010). This can be achieved by a coolant such as water by circulating through a chiller in the laser equipment. The use of water cools the material and removes the debris. In addition, it helps to achieve parallel kerf and multi-directional cutting at high dicing speeds.

Another type of solid-state laser known as “fibre laser” is now becoming popular in metal cutting industry. This laser is produced from a solid gain medium without the use of any liquid or gas. The major advantage of fibre laser is the extremely small spot size. Fibre lasers, with a wavelength of 1.064 μm, can produce an extremely small spot size, which makes it ideal for cutting reflective metals.

4. APPLICATION OF LASER IN APPAREL INDUSTRY

As an all-new process, there are several applications of laser in apparel industry. Laser engraving and cutting technologies now being widely applied in many garment industries, fabric production units, other textile and leather industries (Choudhury and Shirley 2010; Nayak and Khandual 2010). Various applications of laser are discussed in the following section.

4.1 Fabric fault detection

When fabric is received at the stores of a garment production unit, the faults in the fabric can be detected with morphological image processing based on laser (Mallik-Goswami and Datta 2000; Ribolzi et al. 1993; Mursalin et al. 2005). Laser-based optical Fourier transform analysis can be used for fault detection in the fabric as the pattern is repeated at regular intervals. The fabric is focused with a laser and the diffraction gratings obtained from the periodicity of longitudinal and transverse threads in the fabric are superimposed. A Fourier lens is used to produce the diffraction pattern of the fabric. A second Fourier lens with same focal length magnifies and inverts the test sample image. A charge-coupled device (CCD) camera is used to capture the image. The data is transferred and stored in a computer. The computer programming helps in comparing the acquired images with the stored images by converting the image into binary mode. A fault is reported when the measured parameter is deviating from the standard. The severity of the fault depends upon the amount of deviation from the standard.

4.2 Laser cutting

After they were introduced in the 19th century, the fashion designers are widely adopting laser cutting in garment manufacturing (Petrak and Rogale 2001). In synthetic fabrics, laser cutting produces well-finished edges as the laser melts and fuses the edge, which avoids the problem of fraying produced by conventional knife cutters. Furthermore, use of laser cutting is increasingly used for leather due to the precision of cut components. In fashion accessories such as jewellery, laser cutting can be used to produce new and unusual designs to produce a fusion of apparel design and jewellery style.

In laser cutting a laser is used to cut the fabric into the desired pattern shapes. A very fine laser is focused on to the fabric surface, which increases the temperature substantially and cutting takes place due to vaporization. Normally gas lasers (CO_2) are used for cutting of fabric. The cutting machine (Fig. 2) includes a source of laser, a cutting head fitted with mirrors to reflect the laser beam to the cutting line, a computer to control the entire system and a suitable mean for removing the cut parts. The application of inert gases (N_2, He) during cutting prevents the charring and removes debris and smoke from the cutting area. Like the mechanical cutting devices, a laser beam does not become blunt and need sharpening. Automatic single ply laser cutters are faster (30–40 m/min) than automatic multiple ply knife cutters (5–12 m/min). However, while cutting multiple plies, knife cutters are faster per garment cut and also cheaper.

The limitation of laser cutting is the number of lays of the fabric that can be cut by the beam. Best result is obtained while cutting single or a few lays, but the accuracy and precision is not obtained with several plies. In addition there is a chance of the cut edges to be fused together especially in case of synthetics. In some cases the sealing of the edges of cut patterns and sewn garment parts is essential to prevent fraying, where the laser plays the role. As in garment production facilities emphasis is given in multiple lay cutting, the laser cutting seems unlikely to become widespread. However, it is successfully used in cutting of sails where single ply cutting is the norm and a slight fusing of the edge of synthetics and woven materials is desirable. In addition, laser cutting is used in some areas of home furnishing.

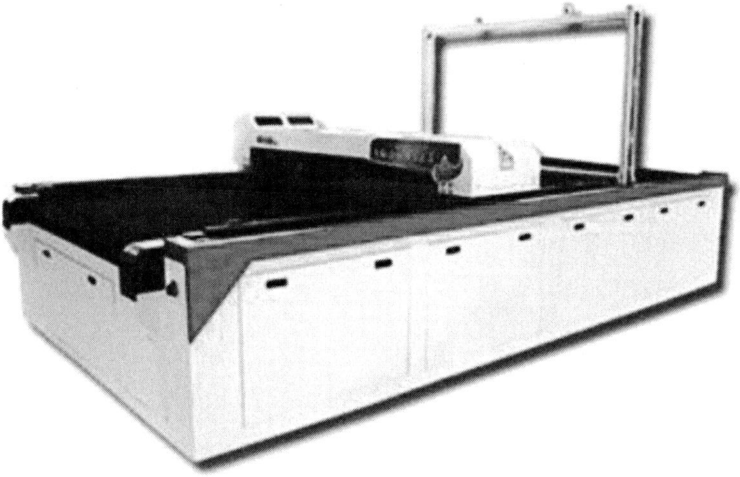

Figure 2. A laser-cutting machine

Laser cutting is cheaper compared with the traditional cutting methods (Mahrle and Beyer2009). Furthermore, as the laser cutting doesn't have mechanical action, high precision of the cut components at high cutting speed are feasible (Mathew et al. 1999). The laser cutters are safer and include simple maintenance features, which can be operated for longer duration. The laser cutters can be integrated to the computer technology. It can produce the products at the same time when designing in the computer. Laser cutting machines have faster speed and simpler operation.

Laser cutting machines are suitable for cutting textile fabrics, composites and lather materials (Caprino and Tagliaferri 1988; Steen et al. 2010; Cenna and Mathew 2002). They can operate for a wide range of fabric, which is not possible with die cutters. Hence, laser cutting machines are gradually been accepted in garment manufacturing. The features of laser applications include:

- Laser marking, laser engraving and laser cutting combined in one step
- No mechanical wear, hence good quality
- No fixation of material is required due to force-free processing
- No fabric fraying in synthetic fibres due to formation of fused edges
- It is clean and lint-free
- Simple process due to integrated computer design

- High quality raw materials and significant cost saving
- Extremely high precision in cutting contours
- High working speed
- Contactless, wear-free technique
- No chips, less waste

5. OBJECTIVE EVALUATION OF SEAM PUCKER

Garment appearance greatly influences garment quality. Seam pucker negatively affects the garment appearance (Nayak and Padhye 2014b, 2015a; Nayak et al. 2010, 2013; Fan and Liu 2000). There are several methods to measure seam pucker, but the conventional rating system developed by American Association of Textile Chemists and Colourists (AATCC) is mainly used. The laser beam can measure the degree of puckering in garments by geometrical models. In this method a seam in the garment is scanned by a 3D laser scanner by putting the garment on a dummy. The laser head can be moved to any 3D space within a confined place by an operator. It is possible to scan the target object from different angles. A pucker profile of the scanned seam can be obtained by processing the image with a 2D digital filter. Physical parameters such as log σ^2 (σ is variance) can be obtained from the pucker profile, which can then be linearly related to grade for seam pucker. From the objectively measured log σ^2, the pucker grade can be objectively evaluated.

6. MASS CUSTOMIZATION

The term mass customization is used when custom-fit garments are obtained depending on the body dimensions and individual's choice. The very first thing to mass customize garments is the accurate measurements of individual's body (Nayak and Padhye et al.2015a). Laser scanning technology is one of the many techniques used for measurement. Laser scanning technology uses one or multiple thin and sharp stripe lasers to measure body size. Cameras are also used to acquire the scene and assist the laser scanner. The body measurements are derived by applying simple geometrical rules (D'Apuzzo 2007; Tong et al. 2012; Ashdown et al. 2004). In order to confirm the harmlessness of the beam, only eye-safe lasers can be used. Additional optical devices such as mirrors can be used to assist a single laser beam. The laser scanning unit (Fig. 3) consisting of light sensors and optical systems focuses on the human body for digitisation. The number of light sensors and optical systems can vary as per the positions of the body. For example, Vitronic1 body scanner consists of three scanning units that can synchronously move vertically along three pillars.

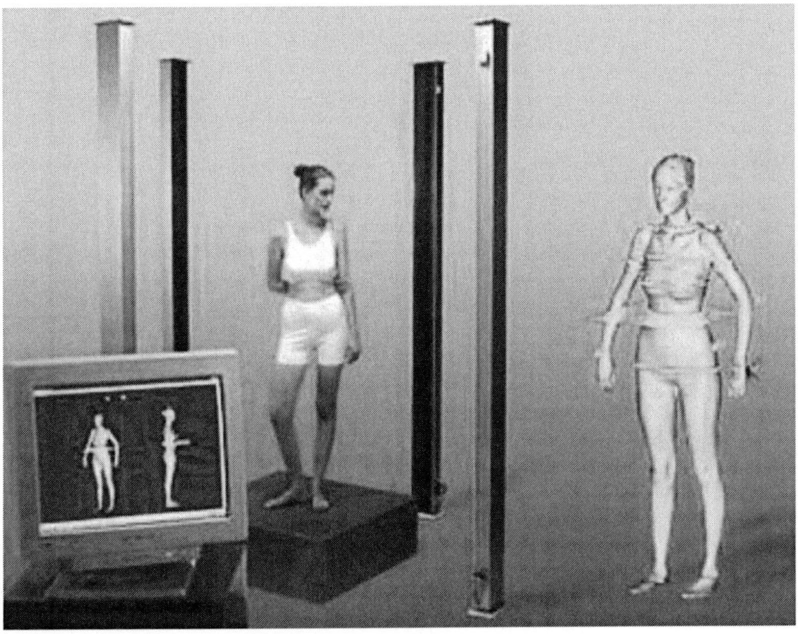

Figure 3. 3D body scanning by Laser scanner

7. LASER-BASED DENIM FADING

Now the age of fading of denim by sandblasting is becoming older as the new technology of laser fading is replacing it (Ortiz-Morales et al. 2003; Tarhan and Sarıışık 2009). In laser fading, a computer drives the laser beam to the material where marking or fading is required. The laser beam decomposes the dye and the resulting vapors are vented away. The material fades only where the beam impacts on the fabric. Commercially two types of lasers are being used: solid based (wavelength of 1 μm) and gas based (wavelength of 10 μm). The desired degree of fading depends upon the wavelength, power density, and pulse width of the laser beam. The method of marking or fading by laser is more environmental friendly as compared to acid washing or sandblasting (Kan et al. 2010). A laser-faded denim sample is shown in the Fig. 4.

Figure 4. Denim faded by laser

8. LASER ENGRAVING

In laser engraving laser is used to mark or engrave an object. The process is very complex, and often computerised systems are used to drive the laser head (Kan et al. 2010; Juciene et al. 2013). In spite of the complexity, very precise and clean engravings can be obtained with high rate of production. The technique does not involve physical contact with the engraving surface, hence, no wear and tear. The marks produced by laser engraving are clean, crisp and permanent. In addition, lasers are faster than other conventional methods used for product imprinting, which provides greater versatility in material selection. One machine can be used to cut through thin materials as well as make engravings on them. Laser engraving is used to engrave the printing screens, for hollowing, for creating pattern buttons, to engrave leather, denim etc. (Fig. 5). Pictures, flower patterns and even personalized signatures can be engraved on leather shoes, leather bag, wallet, leather belt, leather sofa and leather clothes, greatly increasing the added value of products. In addition laser engraving is used to create embroidered pattern in the fabric by colour fading and burning the fabrics. The low cost sealed CO_2 lasers are preferred for laser engraving.

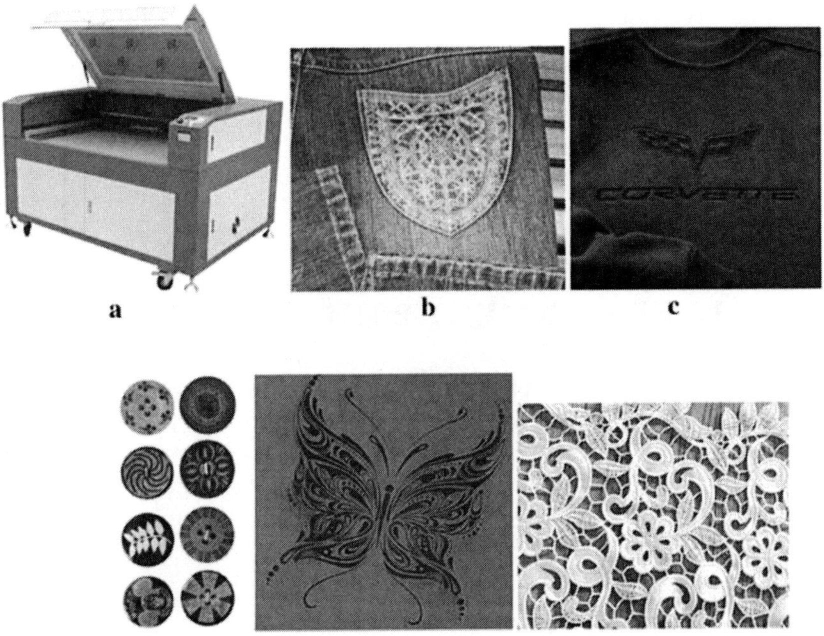

a b c

d e f

Figure 5. Laser engraving items: **a** engraving machine, **b** denim, **c** garment, **d** buttons, **e** leather and **f** embroidery

Denim engraving is another fast-growing application of laser using sealed CO_2 lasers (Juciene et al. 2013; Kan 2014a). The laser is used to create minute designs and patterns on denim fabric as well as finished denims. This technic can be used in place of the traditional techniques such as sandblasting and acid washing. The accuracy and design flexibility is very wide, which can't be achieved by the traditional methods. Lasers can produce 3D effects by techniques such as embroidering, embossing, or even apparent cuts, tears and mends. Any image that is created in a computer aided design (CAD), can be transferred to denim by suitable laser process. While using lasers, features such as good mode quality, high power stability, real-time control of laser power and fast pulse rise-time are the important parameters that can lead to colour change without charring or other damage to the fabric. Such damage could reduce the product life and cosmetically unacceptable. The advantages of laser engraving over traditional methods include:

- High working speed without mechanical contact
- No wear and tear of components
- Reduced waste
- Complete exhaust and filtering
- Exact contours possible

9. WELDED GARMENT PRODUCTION

Welding is an alternative process of joining fabrics for garment production where the thermoplastic materials are joined together by the application of heat. The heat can be supplied by ultrasonic or by high powerful laser (Petrie 2015). The welded garment though weaker than the sewn counterpart, gives better appearance as it does not contain bulky seam and is more flexible.

10. BAR CODE SCANNING

The scanners used to scan the barcodes for product identification typically uses helium–neon (He–Ne) lasers. The laser beam bounces out of a rotating mirror while scanning the code. This sends a modulated beam to a computer, which contains the product information. Semiconductor based-lasers can also be used for this purpose. However, some of the recent manufacturers are using Radio Frequency Identification (RFID) based tags instead of barcodes due to certain advantages (Nayak and Padhye 2014c). The RFID tag can be processed quickly and it avoids the physical handling of the product as in barcode systems (Nayak et al. 2007, Nayak et al. 2015b; Nayak and Padhye 2015b). The mechanism of a bar code scanner is shown in Fig. 6.

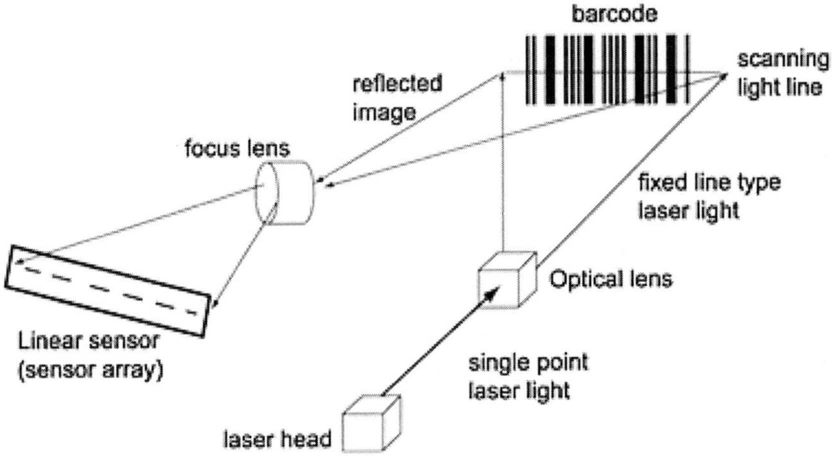

Figure 6. A laser bar code scanner

11. LASER MARKING

Laser can also be used in marking on various surfaces. The advantages of laser marking include fast, high precision and clear marking on products of varying contour and hardness. It can also be used for a wide range of organic polymers where precession can be obtained even with complex designs. Laser marking is durable and can be applied in clothing, leather and metals. Laser marking is considered to be the best choice for branded clothing and marking fashion accessories during processing.

12. MISCELLANEOUS USES

There are many other applications of laser in apparel industry as discussed below:

- In some stitching machines (e.g., automatic welt pocket attaching machine) laser beam is engaged for automatic and accurate positioning of the welt and to make a slit across the fabric.

- In some metal detection machines laser beam is employed to find if any needle part is there in the final finished garment.

- Laser-engraving machine can engrave any decoration on the surface layer of any material, making the products looking high grade and exquisite.

- Laser technology is gaining impetus in garment finishing which can produce various surface ornamentations without any wet processing. This technique is very accurate and can work fast with good repeatability and reproducibility (Hung et al. 2011; Kan et al.2010).

- The application of antimicrobial finishes into textiles has significantly improved compared to the past (Nayak and Padhye 2014a; Nayak et al. 2008). Laser treatment was used for durable antibacterial properties on cotton fabric using silver nanoparticles by Nourbakhsh et al. (Nourbakhsh and Ashjaran 2012).

13. HAZARDS BY LASER AND THEIR CONTROL

13.1 Electrical shock

More people are being killed by electrocution from the laser electronics than blinded from exposure to a laser beam (Sliney 1995). Lethal voltages are present in the power supplies of lasers. If one is not experienced working with high voltages in general and laser power supplies in particular, then the person should not be allowed to carry out any work with the laser. During maintenance, the power supply should be unplugged from its electrical outlet.

13.2 Eye injury potential

Eye is the most vulnerable body part to a laser beam (Henderson et al. 2003; Buckley2010; Barkana and Belkin 2000). With certain lasers severe damage can be caused due to the high concentration of laser energy on retina (Wyrsch et al. 2010). Class 4 lasers can damage the tissues in the eye interior. The class and duration of laser exposure are the deciding factors in the eye injury. For example, no injury is be expected while working with the laser with wavelength in the visible spectrum (400–700 nm). Lasers of Class 3B or Class 4 can lead to an eye injury before the aversion response can protect the eye. Tables 2 and 3 describe various types of laser hazards and precautions to be taken as specified by American National Standards

Institute (ANSI) and Occupational Safety and Health Administration (OSHA).

Table 2. Nature of hazards and precautions for various classes of lasers

Laser class	Nature of hazard (ANSI and OSHA)	Precautions
Class 1	Low power, safe to view	No potential hazard
Class 1M	Low power, hazardous when viewed directly for longer than 1000 s	Do not view directly with optical instruments
Class 2	Low power, hazardous when viewed directly for longer than 0.25 s	Do not stare into the beam
Class 2M	Medium power, nonhazardous when viewed directly for less than 0.25 s	Do not stare into the beam, do not view directly with optical instruments
Class 3A	Medium power (0.5 W), are hazardous when viewed directly	Avoid direct eye exposure
Class 3B	Medium power (0.5 W), are hazardous when viewed directly	Avoid direct eye exposure
Class 4	High power (>0.5 W), produce ocular, skin and fire hazards	Avoid eye or skin exposure to direct or scattered radiation

Table 3. Harmful effect of laser radiation on eye and skin

Laser type	Wavelengths (nm)	Impacts on eye	Impacts on skin
Excimer laser	100–315	Cornea ignition	Sunburn, accelerated aging
He–Ne laser	315–380	Lens opacity	Increased pigmentation
Nd-YAG laser	380–780	Violation of the retina	Darkening of pigment, burns
High performance diode laser	780–1400	Lens opacity, violation of the retina	Burns
CO_2 laser	1400–3000	Lens opacity, burning of the cornea	Burns
CO_2 laser	3000–10,000	Burning of the cornea	Burns

Various standards are devised around the world for the laser related hazards. The standards devised by ANSI are shown in Table 4.

Table 4. Various ANSI standards dealing with laser hazards

Laser standards	Description
ANSI Z136.1 (safe use of lasers)	This standard is the foundation of laser safety programs for industry, military, research and development and higher education (universities)
ANSI Z136.2 (safe use of optical fibre communication systems utilizing laser diode and LED sources)	This standard provides guidelines for the safe use, maintenance, service and installation of optical systems utilizing laser diodes or light emitting diodes operating at wavelengths between 0.6 µm and 1 mm
ANSI Z136.3 (safe use of lasers in health care)	This standard provides guidelines for individuals working with Class 3B and Class 4 lasers and laser systems in health care
ANSI Z136.4 (recommended practice for laser safety measurements for hazard evaluation)	This standard provides guidelines for measurement procedures necessary for the classification and evaluation of optical radiation hazards
ANSI Z136.5 (Safe use of lasers in educational institutions)	This standard provides guideline for organisations and implementation of laser safety and training programs. In addition, it privides graphics for entryway controls, laser installations and laser laboratory layouts
ANSI Z136.6 (safe use of lasers outdoors)	This standard provides guidelines for the safe use of lasers in an outdoor environment such as construction, light shows, scientific, research and military
ANSI Z136.7 (testing and labelling of laser protective equipment)	This standard provides guidelines on the test methods and protocols used to provide eye protection from lasers and laser systems
ANSI Z136.8 (safe use of lasers in research, development, or testing)	This standard provides guidelines on the safe use of lasers and laser systems found in research, development or testing environments, where safety controls common for commercial lasers may either be missing or disabled
ANSI Z136.9 (safe use of lasers in manufacturing environments)	This standard provides guidelines for laser exposures when lasers are used in manufacturing environments. This also includes policies and procedures for safety in both public and private industries as well as product development along with testing

13.3 Skin injury potential

The injury to the skin due to laser radiation is less severe compared to the eye. However, the chances of exposure of skin is higher than that of the eye due to its greater surface area (Yashima et al. 1991; Sliney 1995). The eye injury is more significant than the skin as the loss in the vision is irreparable. In normal laser working condition there is very less chance that a large area of the skin is exposed. The injury to the skin due to laser exposure can be divided into two categories: (a) thermal injury and (b) photochemical induced injury. The former is caused by the high power laser beams, whereas the latter is caused by exposure to scattered ultraviolet (UV) laser radiation (Bernstein et al. 1997). Direct contact or exposure to the laser beam can lead to thermal injuries. Although these injuries are painful, usually they are not serious and proper beam management and hazard awareness can prevent these. In addition, specular or even diffused reflections can lead to photochemical injury over time.

The objective of hazard control methods is primarily stop the laser contacting the skin or entering into the eye (Protection 1996). These control methods can be grouped into three sections such as: (a) administrative controls [labels, signs, standard operating procedures (SOPs), etc.], (b) engineering controls (barriers, blocks, etc.) and (c) protective controls (eyewear, uniform, etc.).

13.4 Administrative controls

The management should only allow the trained persons to work on laser equipment. The operator should follow the instructions as in SOPs. The laser equipment should be switched off while not being used. All laser equipment of Class 3B or Class 4 lasers need to be labelled with “Danger” symbol, specifying the laser class. A Class 3B laser device should mention “LASER RADIATION-AVOID DIRECT EXPOSURE TO THE BEAM”, which must be written above the tail of the sunburst. A Class 4 laser device should mention “LASER RADIATION-AVOID EYE OR SKIN EXPOSURE TO DIRECT OR SCATTERED RADIATION” above the tail of the sunburst.

13.5 Engineering controls

The possibility of accidental exposures to laser hazards can be best controlled by engineering controls. All the Class 3B and Class 4 laser equipment should prevent the access of unauthorized personnel in the working area while laser is operational. All the Class 3B and Class 4 lasers should have a non-flammable cover sufficient to hold the excitation device and the beam. The laser systems can be fitted with key switches or password protected for better safety. In the laser chamber, the setting should guarantee no direct eye contact. The most hazardous aspect of laser use is the beam alignment, where most eye injuries occur. Hence to avoid this, the instructions described in the SOP must be understood. The lowest visible beam power should be used for beam alignment.

13.6 Protective equipment controls

The user of laser equipment should wear appropriate personal protective clothing/device for eye and skin protection during initial setting as well as the normal working (Nayak et al. 2015c). The skin covering and the eyewear protect the skin and eye, respectively from direct exposure.

14. CONCLUSIONS AND FUTURE DIRECTIONS

Laser technology can be used for various applications on materials ranging from metals to textiles with noncontact patterns. In garment production, it can be applied onto different products ranging from home textiles to fashion accessories. In garment manufacturing, CO_2 gas lasers have wide and successful applications. Laser technique, is entirely different from traditional textile processes, as it has the flexibility in design and operation without any pollution or waste material. There are several other advantages of using laser over the conventional methods in cutting, engraving, embossing, denim fading etc. In addition, laser involves lower risk of product damage, use of low consumables and free from disposing of toxic by-products, as there may be with some methods. The laser equipment of today has gradually evolved from those used in early days. The old laser equipment were difficult to run, cumbersome and hard to maintain. However, the modern equipment are easy to operate, simple to learn and easy to maintain. Furthermore, the earlier equipment had several safety issues and they needed replenishment of gasses at regular intervals. The garment production units should take the advantage of applying laser in the post multi-fibre agreement regime to produce more competitive products.

AUTHORS' CONTRIBUTIONS

RN designed the paper, performed the literature survey and completed the paper. RP guided during the paper design and helped in feedback and comments in final format of the paper.

COMPETING INTERESTS

The authors declare that they have no competing interests.

REFERENCES

1. Ashdown, S. P., Loker, S., Schoenfelder, K., & Lyman-Clarke, L. (2004). Using 3D scans for fit analysis. *Journal of Textile and Apparel, Technology and Management, 4*, 1–12.

2. Bahtiyari, M. (2011). Laser modification of polyamide fabrics. *Optics & Laser Technology,43*, 114–118.

3. Barkana, Y., & Belkin, M. (2000). Laser eye injuries. *Survey of Ophthalmology, 44*, 459–478.

4. Belli, R., Miotello, A., Mosaner, P., & Toniutti, L. (2005). Laser cleaning of ancient textiles.*Applied Surface Science, 247*, 369–372.

5. Bernstein, L. J., Kauvar, A. N., Grossman, M. C., & Geronemus, R. G. (1997). The short-and long-term side effects of carbon dioxide laser resurfacing. *Dermatologic Surgery, 23*, 519–525.

6. Buckley, E. (2010). Eye-safety analysis of current laser-based scanned-beam projection systems. *Journal of the Society for Information Display, 18*, 944–951.

7. Caprino, G., & Tagliaferri, V. (1988). Maximum cutting speed in laser cutting of fiber reinforced plastics. *International Journal of Machine Tools and Manufacture, 28*, 389–398.

8. Cenna, A., & Mathew, P. (2002). Analysis and prediction of laser cutting parameters of fibre reinforced plastics (FRP) composite materials. *International Journal of Machine Tools and Manufacture, 42*, 105–113.CrossRef

9. Choudhury, I., & Shirley, S. (2010). Laser cutting of polymeric materials: an experimental investigation. *Optics & Laser Technology, 42*, 503–508.

10. Chow, Y. F., Chan, A., & Kan, C.-W. (2012). Effect of CO2 laser irradiation on the properties of cotton fabric. *Textile Research Journal, 82*, 1220–1234.

11. D'Apuzzo, N. (2007). 3D body scanning technology for fashion and apparel industry. In*Electronic Imaging 2007* (vol. 6491, pp. 64910O–649102). International Society for Optics and Photonics.

12. Dowden, J. (2009). *The Theory of Laser Materials Processing: Heat and Mass Transfer in Modern Technology*. Berlin: Springer Science & Business Media.

13. Fan, J., & Liu, F. (2000). Objective evaluation of garment seams using 3D laser scanning technology. *Textile Research Journal, 70*, 1025–1030.

14. Henderson, R., & Schulmeister, K. (2003). *Laser safety*. Boca Raton: CRC Press.

15. Hung, O., Song, L., Chan, C., Kan, C., & Yuen, C. (2011). Using artificial neural network to predict colour properties of laser-treated 100% cotton fabric. *Fibers and Polymers, 12*, 1069–1076.

16. Istook, C. L., & Hwang, S.-J. (2001). 3D body scanning systems with application to the apparel industry. *Journal of Fashion Marketing and Management, 5*, 120–132.

17. Juciene, M., Urbelis, V., Juchnevičienė, Ž., & Čepukonė, L. (2013). The effect of laser technological parameters on the color and structure of denim fabric. *Textile Research Journal.*

18. Kan, C.-W. (2014a). CO_2 laser treatment as a clean process for treating denim fabric.*Journal of Cleaner Production, 66*, 624–631.

19. Kan, C. (2014b). Colour fading effect of indigo-dyed cotton denim fabric by CO_2 laser.*Fibers and Polymers, 15*, 426–429.

20. Kan, C. (2015). *Washing techniques for denim jeans* (p. 313). Denim: Manufacture, Finishing and Applications.

21. Kan, C., Yuen, C., & Cheng, C. (2010). Technical study of the effect of CO_2 laser surface engraving on the colour properties of denim fabric. *Coloration Technology, 126*, 365–371.

22. Kovacs, L., Zimmermann, A., Brockmann, G., Gühring, M., Baurecht, H., Papadopulos, N., et al. (2006). Three-dimensional recording of the

human face with a 3D laser scanner.*Journal of Plastic, Reconstructive and Aesthetic Surgery, 59*, 1193–1202.

23. Lu, J.-M., Wang, M.-J. J., Chen, C.-W., & Wu, J.-H. (2010). The development of an intelligent system for customized clothing making. *Expert Systems with Applications, 37*, 799–803.

24. Lucas, J., Belino, N., Miguel, R., Pereira, M., & Ribeiro, L. (2015). *Digital printing techniques for denim jeans* (p. 287). Denim: Manufacture, Finishing and Applications.

25. Mahrle, A., & Beyer, E. (2009). Theoretical aspects of fibre laser cutting. *Journal of Physics. D. Applied Physics, 42*, 175507.

26. Mallik-Goswami, B., & Datta, A. K. (2000). Detecting defects in fabric with laser-based morphological image processing. *Textile Research Journal, 70*, 758–762.

27. Mathew, J., Goswami, G., Ramakrishnan, N., & Naik, N. (1999). Parametric studies on pulsed Nd: YAG laser cutting of carbon fibre reinforced plastic composites. *Journal of Materials Processing Technology, 89*, 198–203.

28. Mursalin, T. E., Eishita, F. Z., & Islam, R. (2005). Fabric defect inspection system using neural network and microcontroller. *Statistics, 50*, 100.

29. Nayak, R., Chatterjee, K., Khandual, A., & Jajpura, L. (2008). Evaluation of functional finishes-An overview. *Man-Made Textiles in India, 51*, 131–135.

30. Nayak, R., Chatterjee, K., Khurana, G., & Khandual, A. (2007). RFID: tagging the new era.*Man-Made Textiles in India, 50*, 174–177.

31. Nayak, R., & Khandual, A. (2010). Application of laser in apparel industry. *Colourage, 57*, 85–90.

32. Nayak, R., & Padhye, R. (2014a). Antimicrobial finishes for textiles. In R. Paul (Ed.),*Functional Finishes for Textiles: Improving Comfort, Performance and Protection*. Amsterdam: Elsevier.

33. Nayak, R., & Padhye, R. (2014b). The care of apparel products. In R. Sinclair (Ed.),*Textiles and fashion: Materials, design and technology*. United Kingdom: Elsevier.

34. Nayak, R., & Padhye, R. (2014c). Introduction: the apparel industry. In R. Nayak & R. Padhye (Eds.), *Garment Manufacturing Technology*. Amsterdam: Elsevier.

35. Nayak, R. & Padhye, R. (2015a). Care labeling of clothing. In R. Nayak. & R. Padhye (Eds.),*Garment Manufacturing Technology.*Amsterdam: Elsevier

36. Nayak, R., & Padhye, R. (2015b). *Garment Manufacturing Technology*. Amsterdam: Elsevier.

37. Nayak, R., Padhye, R., Dhamija, S., & Kumar, V. (2013). Sewability of air-jet textured sewing threads in denim. *Journal of Textile and Apparel, Technology and Management,8*, 1–11.

38. Nayak, R., Padhye, R., & Gon, D. P. (2010). Sewing performance of stretch denim. *Journal of Textile and Apparel, Technology and Management, 6*(3), 1–9.

39. Nayak, R., Padhye, R., Wang, L., Chatterjee, K., Gupta, S. (2015a). The role of mass customisation in the apparel industry. *International Journal of Fashion Design, Technology and Education, 8*, 162–172.

40. Nayak, R., Singh, A., Padhye, R., & Wang, L. (2015b). RFID in textile and clothing manufacturing: technology and challenges. *Fashion and Textiles, 2*, 1–16.

41. Nayak, R., Padhye, R., & Wang, L. (2015c). How to dress at work. In S. Patole (Ed.),*Management and Leadership–A Guide for Clinical Professionals*, Germany: Springer

42. Nourbakhsh, S., & Ashjaran, A. (2012). Laser treatment of cotton fabric for durable antibacterial properties of silver nanoparticles. *Materials, 5*, 1247–1257.

43. Ondogan, Z., Pamuk, O., Ondogan, E. N., & Ozguney, A. (2005). Improving the appearance of all textile products from clothing to home textile using laser technology.*Optics and Laser Technology, 37*, 631–637.

44. Ortiz-Morales, M. N., Poterasu, M., Acosta-Ortiz, S. E., Compean, I., & Hernandez-Alvarado, M. R. (2003). A comparison between characteristics of various laser-based denim fading processes. *Optics and Lasers in Engineering, 39*, 15–24.

45. Ozguney, A. T. (2007). The comparison of laser surface designing and pigment printing methods for the product quality. *Optics and Laser Technology, 39*, 1054–1058.

46. Petrak, S., & Rogale, D. (2001). Methods of automatic computerised cutting pattern construction. *International Journal of Clothing Science and Technology, 13*, 228–239.

47. Petrie, E. (2015). Alternative fabric-joining technologies 13. In R. Nayak, R. Padhye (Eds.),*Garment Manufacturing Technology.* Cambridge, UK: Elsevier.

48. Potluri, P., & Atkinson, J. (2003). Automated manufacture of composites: handling, measurement of properties and lay-up simulations. *Composites Part A Applied Science and Manufacturing, 34*, 493–501.

49. Powell, J. (1993). *CO2 laser cutting*. Cambridge: Univ Press.

50. Protection, I. C. O. N.-I. R. (1996). Guidelines on limits of exposure to laser radiation of wavelengths between 180 nm and 1,000 [mu] m. *Health Physics, 71*, 804–819.

51. Ready, J. F., Farson, D. F., & Feeley, T. (2001). *LIA handbook of laser materials processing*. Orlando: Laser Institute of America.

52. Ribolzi, S., Merckle, J., Gresser, J., & Exbrayat, P. (1993). Real-time fault detection on textiles using opto-electronic processing. *Textile Research Journal, 63*, 61–71.

53. Schuocker, D. (1989). Laser cutting. *Material and Manufacturing Process, 4*, 311–330.

54. Simmons, K. P., & Istook, C. L. (2003). Body measurement techniques: comparing 3D body-scanning and anthropometric methods for apparel applications. *Journal of Fashion Marketing and Management, 7*, 306–332.

55. Sliney, D. H. (1995). Laser safety. *Lasers in Surgery and Medicine, 16*, 215–225.

56. Steen, W., Watkins, K. G., & Mazumder, J. (2010). *Laser material processing*. Amsterdam: Springer Science & Business Media.

57. Sutcliffe, H., Cooper, M., & Farnsworth, J. (2000). An initial investigation into the cleaning of new and naturally aged cotton textiles using laser radiation. *Journal of Cultural Heritage, 1*, S241–S246.

58. Tarhan, M., & Sarıışık, M., (2009). A comparison among performance characteristics of various denim fading processes. *Textile Research Journal, 79*, 301–309.

59. Thawari, G., Sundar, J. S., Sundararajan, G., & Joshi, S. (2005). Influence of process parameters during pulsed Nd: yAG laser cutting of nickel-base superalloys. *Journal of Materials Processing Technology, 170*, 229–239.

60. Tong, J., Zhou, J., Liu, L., Pan, Z., & Yan, H. (2012). Scanning 3d full human bodies using kinects. *Visualization and Computer Graphics, IEEE Transactions on, 18*, 643–650.

61. Vilumsone-nemes, I. (2012). *Industrial Cutting of Textile Materials.* Amsterdam: Elsevier.

62. Wyrsch, S., Baenninger, P. B., & Schmid, M. K. (2010). Retinal injuries from a handheld laser pointer. *New England Journal of Medicine, 363*, 1089–1091.

63. Yashima, Y., McAuliffe, D. J., Jacques, S. L., & Flotte, T. J. (1991). Laser-induced photoacoustic injury of skin: effect of inertial confinement. *Lasers in Surgery and Medicine, 11*, 62–68.

64. Yuan, G. X., Jiang, S. X., Newton, E., Fan, J. T., & Au, W. M. (2012). Application of laser treatment for fashion design. *Journal of the Textile Institute, 103*, 48–54.

CHAPTER 3

Medical Smart Textiles Based on Fiber Optic Technology: An Overview

Carlo Massaroni, Paola Saccomandi and Emiliano Schena *

Center for Integrated Research, Università campus Bio-Medico, Alvaro del Portillo, 21, Rome 00128, Italy

ABSTRACT

The growing interest in the development of smart textiles for medical applications is driven by the aim to increase the mobility of patients who need a continuous monitoring of such physiological parameters. At the same time, the use of fiber optic sensors (FOSs) is gaining large acceptance as an alternative to traditional electrical and mechanical sensors for the monitoring of thermal and mechanical parameters. The potential impact of FOSs is related to their good metrological properties, their small size and their flexibility, as well as to their immunity from electromagnetic field. Their main advantage is the possibility to use textile based on fiber optic in a magnetic resonance imaging environment, where standard electronic sensors cannot be employed. This last feature makes FOSs suitable for monitoring biological parameters (e.g., respiratory and heartbeat monitoring) during magnetic resonance procedures. Research interest in combining FOSs and textiles into a single structure to develop wearable sensors is rapidly growing. In this review we provide an overview of the state-of-the-art of textiles, which use FOSs for monitoring of mechanical parameters of physiological interest. In particular we briefly describe the working principle of FOSs employed in this field and their relevant advantages and disadvantages. Also reviewed are their applications for the monitoring of mechanical parameters of physiological interest.

KEYWORDS

smart textiles; fiber optic sensors; fiber Bragg grating sensors; respiratory monitoring; macrobending sensors; hetero-core fiber optics; Magnetic Resonance Imaging; MR-compatibility

1. INTRODUCTION

The growing interest in smart textiles for medical applications is driven by the aim to increase the mobility of patients who need a continuous monitoring of physiological parameters [1,2,3,4]. Smart textiles are able to interact with the environment; therefore they embed one or more sensors to monitor various mechanical, thermal and chemical parameters (e.g., strain, temperature, displacement, oxygen blood saturation) [5,6,7]. During the last decades the use of fiber optic-based sensors (FOSs) has been gaining acceptance in a large number of applications in the fields of civil engineering, the automotive industry and medicine [8,9,10], among others. These sensors allow the measurement of physical and chemical parameters employing a large number of working principles and configurations [4]. FOSs can be divided into intrinsic sensors, where the fiber optic constitutes the sensing element, and extrinsic ones, where the fiber optic is only used as a medium to transport light. There are a number of reviews and books that focus on the description of FOSs [11,12,13,14,15], hence it is not possible to describe all the applications and working principles of FOSs in one journal article.

FOSs have good metrological properties (*i.e.*, low zero drift and sensitivity drift, good accuracy and good sensitivity and large bandwidth), they offer the possibility to implement distributed sensors and they are immune to electromagnetic interferences. These features make FOSs an emerging solution for the monitoring of physiological parameters and more generally for applications in medicine [9]. In fact, thanks to the mentioned valuable characteristics, FOSs can compete with other traditional electrical and mechanical sensors, and in many fields of application the superiority of their performances has been demonstrated. Lastly, the possibility to develop Magnetic Resonance (MR)-compatible sensors further motivates the growing interest in this technology.

Fiber optic technology is particularly attractive for application in smart textiles because it allows both sensing and signal transmission. Moreover, polymer optical fibers (POFs) are cheap, lightweight, flexible and robust, and they are able to measure high strain values without damage.

During the last decade, several groups of research have focused their efforts on obtaining a substantial development in the integration of smart textiles and fiber optic technology [16,17] as reported in a recent review [18].

In this paper, we aim at describing the use of smart textiles in medicine employing FOSs; in particular we focus on their use for monitoring physiological parameters by measuring mechanical variables and on the metrological properties and performances of the devices reported in literature. Also reviewed are the working principles of FOSs most frequently used in smart textiles for the mentioned applications. In Section 2 the working principles of the FOSs used in smart textiles and their applications in medicine are described. In Section 3 the applications in physiological monitoring of smart textiles based on FOSs along with their main advantages and drawbacks are reviewed.

2. WORKING PRINCIPLE OF FIBER OPTIC SENSORS USED IN SMART TEXTILES

FOSs can be designed using a large number of working principles. In this section the ones used to develop smart textiles for measuring mechanical variables in physiological monitoring are reviewed. In particular we focus on sensors based on the fiber Bragg grating technology and on intensity-based FOSs. The next subsections describe the working principle of these sensors with a brief description of their application in medicine.

2.1. Fiber Bragg Grating Sensors

Fiber Bragg grating (FBG) sensors consist of a periodic perturbation of the refractive index along the fiber core length obtained by exposure of the core to an intense optical interference pattern. Hill reported the first fabrication of FBGs in the late 1970s [19], but a major breakthrough occurred a decade later, when Meltz *et al.* improved the technique for their fabrication [20]. This last study can be considered the milestone, which enabled the development of FBG sensors for a large number of applications.

Basically, an FBG can be considered as a short segment of a fiber optic (usually FBG longer than 3 mm–6 mm are employed, although for particular application smaller FBG are required), which reflects a narrow range of wavelengths and transmits all others. A schematic representation of the working principle of an FBG sensor and of its response to strain is shown in Figure 1.

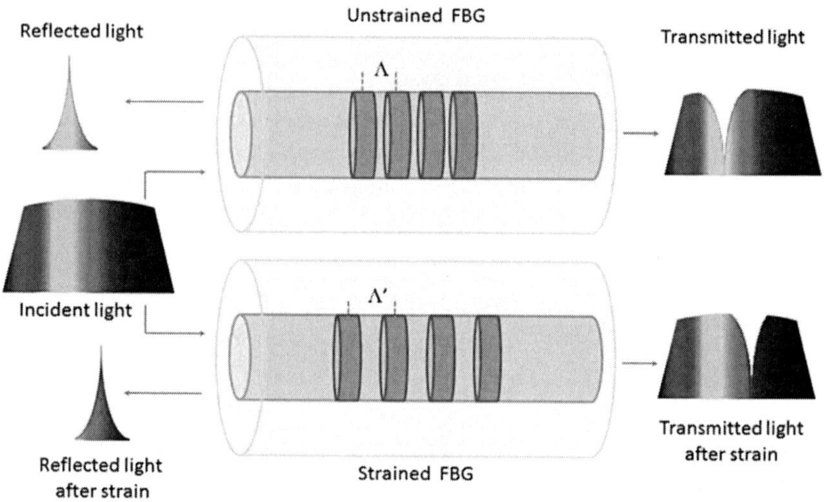

Figure 1. Schematic of the working principle of Fiber Bragg grating (FBG) sensors, and its response to strain.

The wavelength of the input light that is back-reflected (λ_B) is sensitive to temperature and strain. In fact, λ_B can be expressed by the following equation, which is the first-order Bragg condition of the grating:

$$\lambda B=2\cdot\Lambda\cdot\eta eff$$

(1) where Λ is the spatial period of the grating and η_{eff} is the effective refractive index of the fiber core. Both the terms are sensitive to strain and temperature, as a consequence, the use of a proper configuration allows estimating strain and temperature or both, by monitoring changes of λ_B. The dependence of λ_B can be described by the fractional changes of Λ and η_{eff}:

$$\Delta\lambda B\lambda B=\Delta\Lambda+\Delta\eta eff$$

(2) Actually, due to the different sensitivity of Λ and η_{eff} to temperature changes and strain, the FBGs sensors used to estimate strain are based on the λ_B shift due to Λ changes, the FBG sensors used to monitor temperature are based on the $\lambda_B$shift due to η_{eff} changes. Equation (1) shows an important advantage of the FBG: their output is not affected by fluctuation of source intensity.

FBG sensors are employed in a large number of industrial fields to monitor different physical variables, including, among others, temperature, pressure, flow and vibrations. Moreover, FBGs have been largely used to monitor physiological parameters and more in general in medical fields, such as stroke volume, blood pressure and heartbeats [21,22], in microsurgery [23,24], foot pressure in diabetic patients [25], in biomechanical studies [26,27], in the monitoring of temperature during thermal ablation of cancer [28,29,30,31], in respiratory monitoring system [32,33], and in tactile sensing [34,35].

FBGs sensors are considered to hold great potential for application in medicine due to their good metrological properties. Moreover their characteristics of biocompatibility, non-toxicity and chemical inertness, as well as their small size and flexibility make them particularly attractive for invasive measurements during *in vivo* trials [36]. Lastly, they are suitable for application in environments with high electromagnetic noise, thanks to their immunity to electromagnetic interferences and their intrinsic MR-compatibility [37].

2.2. Intensity Modulated FOSs

Intensity-modulated FOSs modulate light intensity, measured by a secondary element (e.g., a photodiode), in response to an environmental effect. A simple configuration of this kind of sensor is shown in Figure 2A. Two optical fibers are held in close proximity to each other; the light is injected into one of the optical fibers; as the light expands into a cone of light, its intensity, emitted by the first fiber and conveyed into the other one, depends on the distance (*d*) between the two fiber tips. Therefore, the light intensity can be considered an indirect measurement of the distance between the two fibers and of other physical variables influencing this distance. A similar configuration can be designed either by using a single fiber and a mirror (Figure 2B) or by using more than one fiber to obtain a differential configuration. The differential solution allows neglecting the influence of the light source intensity on sensor output. Another configuration to develop intensity-modulated FOSs is underpinned by the phenomenon that the light is lost from an optical fiber when it is bent. In particular a bent radius causes a leakage of the light traveling within the core of the fiber into the cladding with a resulting intensity modulation of light propagating through an optical fiber (Figure 2C). Macrobending sensors based on hetero-core fibers have been proposed to measure several physical properties [38,39].

Intensity-modulated FOSs have been used in different medical fields starting from the late 1960s, when Lekholm and Lindstrom proposed a sensor for intravascular pressure monitoring [40]. A similar sensor was proposed to monitor the intracranial pressure [41], and commercially available intracranial pressure intensity-modulated FOSs are produced by Camino Laboratories Inc. Their performances are largely investigated in clinical settings [42]. Moreover these sensors are used for pressure and temperature monitoring [43].

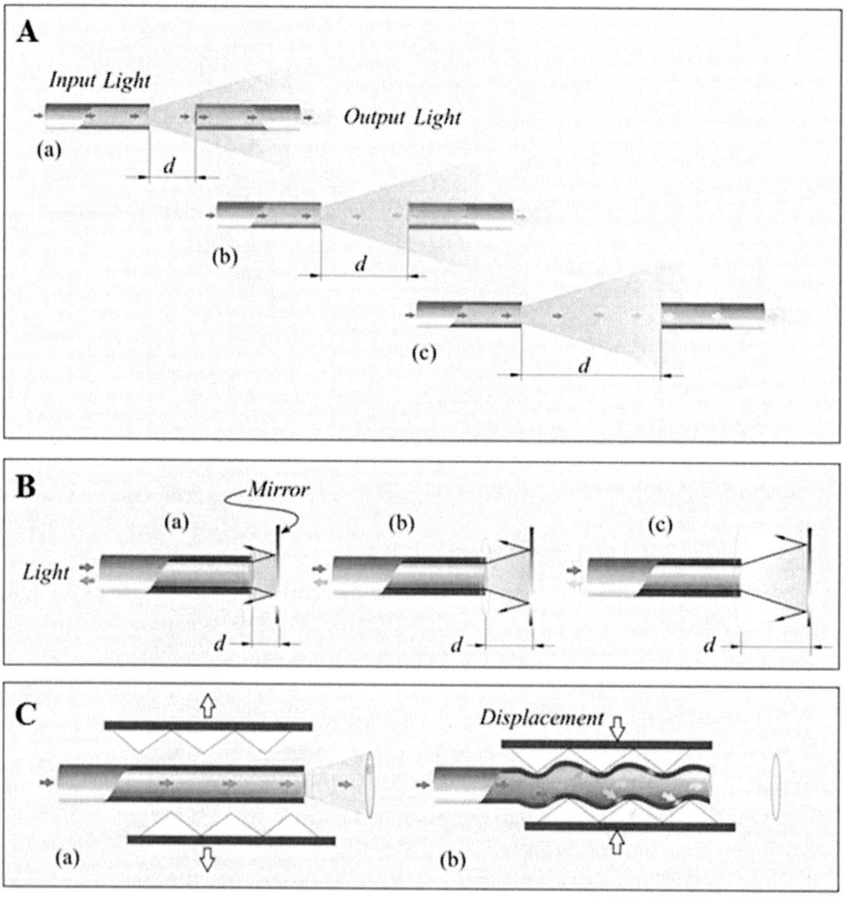

Figure 2. (**A**) Schematic of the working principle of an intensity modulated sensor using two fiber optic; (**B**) schematic of the working principle of an intensity modulated sensor using a fiber optic and a mirror; (**C**) schematic of the light lost from the fiber core caused by bending (adapted from [10]).

Macrobending FOSs find applications in medicine mainly in the monitoring of respiratory movements [44,45,46]. These sensors are largely used in smart textiles, therefore their medical applications will be described in more detail in the following section. A particular approach based on the bending of the fiber is also used to develop flow sensors for mechanical ventilation [47,48].

3. SMART TEXTILES BASED ON FIBER OPTIC SENSORS: MEDICAL APPLICATIONS

Smart technical textiles are by definition textiles that can interact with their environment. Their ability to sense physical and chemical parameters can be accessed by embedding several kinds of sensors. The use of FOSs on textile is particularly attractive in some medical fields because of the possibility to use this technology during MR procedures, the cost reduction of key optical components and the improvement of the component quality, as well as the good metrological properties of these sensors. In particular, polymer optical fibers (POFs) match well with the requirements for application in smart textiles, being cost effective, lightweight and robust; moreover POFs are able to measure high strain values of several ten percent without fiber damage [49].

Significant developments of the integration of FOSs in textiles were driven by several groups of research, and in particular by the group involved in the European project OFSETH (optical fiber sensors embedded into technical textile for healthcare). OFSETH aimed at investigating the possibility to use FOSs embedded into textiles to monitor several physiological parameters.

In the following two subsections the application of smart textiles based on FBG sensors and on intensity modulated FOSs are described.

3.1. Smart Textiles Based on FBG Sensors: Medical Applications

FBG is one of the most frequently employed technologies to design smart textiles based on fiber optics, thanks to their good sensitivity to strain. This characteristic allows developing several configurations based on the measurement of the strain experienced by the FBG sensor to monitor different parameters of physiological interest.

In particular this solution has been employed to monitor respiratory movements. The use of FBG to monitor respiratory movements and breathing rate has been demonstrated in the past [32,50,51], but only during the last decade have they been embedded in smart textiles. A number of studies regarding this topic have been proposed by the groups involved in the OFSETH project. A smart textile embedding two different FOSs (*i.e.*, FBG and macrobending) for respiratory monitoring has been developed. FOSs and the use of MR-compatible connectors allow the use of the proposed smart textiles on anesthetized patients during MR procedures. Indeed, these sensors are free from metallic or electrical conductive wires (when using a custom made MR-compatible connector); in addition, they are remotely interrogated via an optical fiber cable allowing the location of the monitoring unit outside of the MR field [52,53], as schematically reported in Figure 3.

The high FBG sensitivity to strain (*i.e.*, $\approx 1.2\ \text{pm}\cdot\mu\varepsilon^{-1}$) allows discriminating small strains; therefore they have been used to monitor thoracic movements that are smaller than abdominal ones. Two different methods to embed the FBG sensors within textiles (stitching and crochet) have been proposed. The calibration of the system, which embeds the FBG by stitching (seeFigure 4A), has been performed by stretching the textiles in steps of 0.4% up to 40%. During the calibration, the FBG experiences strains up to 0.8%. The system shows good linearity from about 8.5% to 35%–40% of textile stretching, with a sensitivity of 0.35 nm/% and an accuracy better than 0.1% of elongation. Only preliminary experiments on the textiles developed with the crochet method have been performed. The integration of this sensor with a different FOS has been assessed on ten healthy volunteers [54]. Trials to evaluate the long-term properties and the stability of the respiratory sensor by an *ad hoc* developed simulator have been performed as well [55]. In particular the mechanical robustness of the sensor has been tested with more than 90,000 cycles in 129 hours with a simulated breathing rate between 10 breaths per minute and 12 breaths per minute [56]. A further valuable characteristic of the proposed system (see Figure 4B) is that it enables the continuous measurement of respiratory movement providing free access to all vital organs for medical staff actions [57].

A recent interesting study proposed a simple wearable system based on a single FBG sensor, which allows the simultaneous detection of both heart rate and respiratory cycles [58]. The main innovation is related to the structure in which the FBG sensor is embedded, which is [9] a PVC laminate resulting in a strain-sensitive foil, manufactured by an industrial spread-coating process. This integrated solution shows a sensitivity of $0.8\ \text{pm}\cdot\mu\varepsilon^{-1}$. The authors performed trials on healthy volunteers, using two filters for breathing rate monitoring (band-pass filter tuned in the range 0.1–0.4 Hz) and cardiac frequency monitoring (band-pass in the range 0.5–1.3 Hz). The same group of research developed a system for breathing rate and cardiac frequency monitoring able to work in a wide range of temperature [59]. The authors selected the polychloroethanediyl as carrier material due to its excellent performance/cost and to the good sensitivity to strain. The system showed a linear response for elongation ranging from 0.6% to 1.6%. In this range the FBG output experienced an increase of about 8 nm; therefore the sensitivity was about 8 nm/%. They also investigated the sensors' output changes with changes in temperature and found an increase of Bragg wavelength of about 1.5 nm for a temperature increase of 140 °C (a thermal sensitivity of about 10.7×10^{-3} nm/°C).

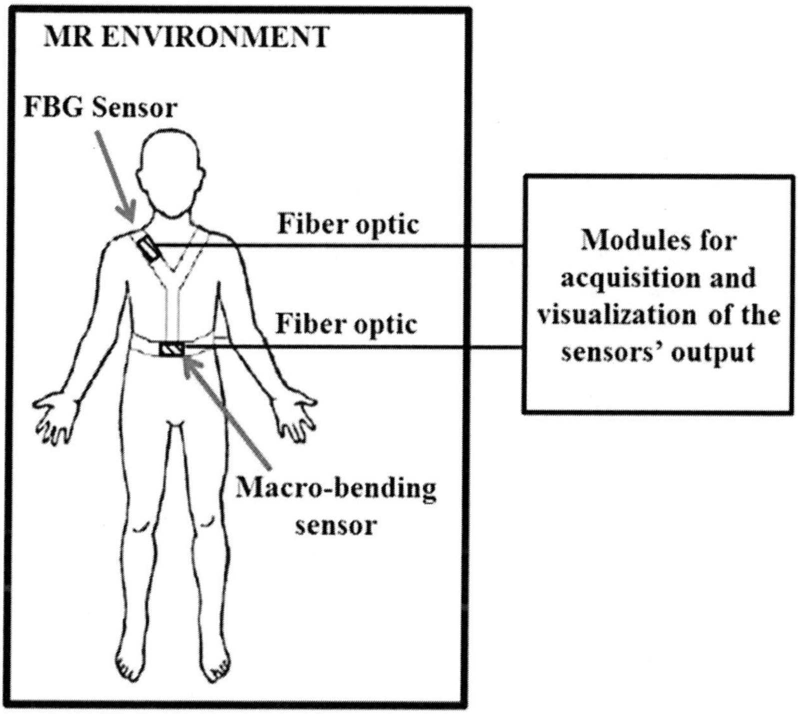

Figure 3. Schematic representation of the monitoring systems proposed by and developed in the optical fiber sensors embedded into technical textile for healthcare (OFSETH) project.

Other simple solutions have been proposed to monitor respiration and heart activity. In particular, Dziuda and coauthors developed a system which consisted of a Polymethyl methacrylate (PMMA) board with the size of 220 × 95 mm^2 and a thickness of 1.5 mm. An FBG sensor was attached with epoxy adhesive [60]. They experimentally assessed the error of the system in the measurement of breaths per minute and heartbeats per minute during magnetic resonance imaging examinations, showing promising results (about one breath per minute and about three heartbeats per minute). The same group of researchers reported a system based on two FBG sensors positioned orthogonally to each other [61]. This group validated their system on three patients during MRI procedures by comparing their results with the ones obtained by an MRI-compatible portable module [62]. The results were promising (a relative error lower than 8% can be considered satisfactory considering that the system is intended for monitoring rather than diagnosis). Recently, they proposed a system for heart rate monitoring. It is MR-compatible and has been tested on seven volunteers showing a root mean square error of less than six beats per minute [63].

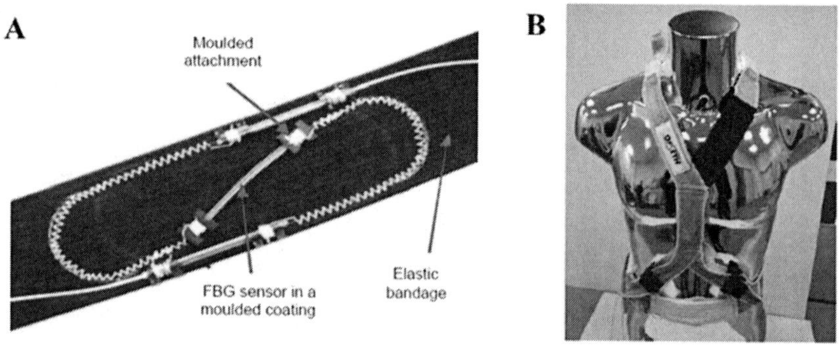

Figure 4. (**A**) Design of the FBG sensor developed in OFSETH project (adapted from [57]); (**B**) MRI-compatible sensing harness which embeds the fiber optic sensors for respiratory monitoring (adapted from [57]).

Allsop and coauthors developed a wearable system for respiratory function monitoring [64]. Their system employs an array of 40 in-line FBG sensors that produce 20 curvature sensors at different locations, each sensor consisting of two FBGs. They carried out experiments to measure the absolute volumetric changes of the human torso and estimated an error of 6% on the average volume.

Lastly, Li and coworkers developed a wearable sensor in intelligent clothing based on FBG for human body temperature monitoring [65]. They partly embedded an FBG in a polymer filled strip to improve the sensitivity of the measuring system. This way they obtained a temperature sensitivity of 150 pm·°C^{-1}, almost 15 times higher than that of the bare FBG. To measure human body temperature, they distributed five FBGs in five places (*i.e.*, left chest, right chest, left armpit, right armpit and at center of the upper back).

Moreover, they developed a model to estimate the body temperature by the data of the five FBGs. With this method they found an accuracy of ±0.1 °C.

The main characteristics of the abovementioned smart textiles and their applications are reported in Table 1.

3.2. Smart Textiles Based on Intensity-Modulated FOSs

FOSs based on intensity modulation and in particular macrobending sensors are employed to develop smart textiles for monitoring of physiological parameters.

Textiles based on these kinds of FOSs were proposed by the groups of research involved in the OFSETH project. In particular, they developed smart textiles based on macrobending FOS for the monitoring of respiratory rate on patients during MRI procedures [52,53]. The sensors are MR-compatible and the monitoring devices are located out of the magnetic resonance environment (see Figure 3). A standard single mode fiber was embedded within a textile as shown in Figure 5A. This macrobending sensor, due to the lower sensitivity than the FBG sensors, was used to monitor abdominal movements, because these movements are larger than thoracic ones (Figure 5B). The authors found large oscillations in the sensor's output during the stretching of the textile, which could cause a wrong computation of the breathing rate. As a consequence, they designed a sensor with more periods of the U-shape on the textile bandage (Figure 5B). This solution allowed to substantially increase the sensitivity (e.g., for a textile elongation of 38%, the sensor response increased from less than 3 dB with a single-loop design to more than 28 dB with a 10 loops-design). Lastly they developed a sensor based on the optical time-domain reflectometry (OTDR) in polymer optical fiber (POF). Basically, the macrobending entails a change of the backscattering in POF that can be sensed by the technique of OTDR [56]. This solution allows measuring strain in different locations of a single fiber (distributed measurement). A respiratory simulator was employed to test the robustness of both the macrobending sensor and the sensor based on the OTDR technique. Both these sensors show low variations after cycles at 10 breaths per minute and elongation up to 3% and 5% for the OTDR sensor and for the macrobending one, respectively. These sensors were also tested on healthy volunteers.

Table 1. Smart textiles and wearable systems based on fiber optic sensors: working principle, medical application and metrological properties.

Reference	Working Principle	Medical Application	Metrological Properties and Other Features
[52,53,56]	Silica FBGs	Respiratory monitoring during MRI procedures	Non-invasive; MR compatible; good linearity in a wide range of strains with sensitivity = 0.35 nm/%; accuracy better than 0.1% of elongation
[58,59]	Silica FBGs	Cardiac and Respiratory monitoring during MRI procedures	Non-invasive; MR compatible; sensitivity of 0.8 nm/$\mu\varepsilon^{-1}$
[59]	Silica FBGs	Cardiac and Respiratory monitoring	Non-invasive; Sensitivity of 8 nm/%; good linear trend; thermal sensitivity $\approx 10.7 \times 10^{-3}$ nm/°C
[60,62]	Silica FBGs	Cardiac and Respiratory monitoring during MRI procedures	Non-invasive; MR compatible; Simple design; Good accuracy in terms of breathing rate (±1 bpm) and heartbeat per minute (±3 bpm); relative error in patients during MRI procedures <8%
[61]	Silica FBGs	Cardiac and Respiratory monitoring during MRI	Non-invasive; Simple design; Flat frequency response in the range of interest (0.5 Hz up to 20 Hz); maximum relative error of 12%
[63]	Silica FBG	Heart rate monitoring	Non-invasive; MR compatible; Root mean square error lower than 6 beats per minute
[64]	Silica FBGs	Respiratory function monitoring	Non-invasive; 6% of error on the average volume
[65]	Bare FBG	Body temperature monitoring	Non-invasive; Sensitivity of 150 pm/°C in the range of interest (from 33 °C to 42 °C); accuracy 0.1 °C

[52,53,56]	Macro-bending/OTDR technique	Respiratory monitoring	Non-invasive; MR compatible; Good sensitivity stability after 172800 cycles (variations < 10%); low cost component for their interrogation
[66]	Intensity modulated	Respiratory monitoring	Non-invasive; MR compatible; low cost component
[67,68]	Intensity modulated	Respiratory monitoring	Non-invasive; low cost component
[69,70]	Macrobending hetero-core fiber optic	Respiratory monitoring	Non-invasive; low cost component; good agreement with the breathing rate measured by a commercial device
[71]	microbending	Respiratory monitoring during MRI procedures	Non-invasive; MR compatible; Accuracy better than ±2 breaths per minute
[72]	microbending	Respiratory rate and heart rate	Non-invasive; MR compatible; Accuracy better than ±2 breaths or beats per minute for respiratory monitoring heart rate
[73]	microbending	Heartbeat and respiratory monitoring	Non-invasive; low cost component; good agreement with the heart beat measured by a commercial device

An interesting solution to develop an MR-compatible sensor wearable sensor for respiratory monitoring based on intensity modulation was presented in [66]. It consists of a PMMA tube, a mirror, a spring and a plastic optical fiber. Abdominal movement causes a variation of the distance between the mirror and the distal end of the plastic optical fiber that is related to the intensity of reflected light coupled to the fiber (as shown in the schematic representation in Figure 2B). The authors also tested the sensor during an MR procedure, and they did not find any negative effects related to patient safety and image quality.

Krehel and coauthors developed a textile for respiratory monitoring based on FOS previously described [67,68]. Basically, the working principle of the sensor employed can be explained as follows: the fiber optic geometry changes when a force (or pressure) is applied on the fiber; these geometry changes affect the wave guiding properties and hence induce light loss in the optical fiber. They characterized the sensor in [67], showing a range of measurement for force applied on 3 cm of fiber length up to 40 N, with a discrimination threshold of 0.05 N. They also performed the feasibility assessment of the wearable system for breathing rate monitoring [68]. The trials were performed at two breathing rates and at different positions of the sensing textile on the human body. Lastly, the comparison with a commercial device for respiratory measurements was performed using the Bland Altman analysis. The results showed that a large part of the differences between the measurements obtained by the FOS textile and the commercial device was concentrated in the range ±3 min^{-1}, and the limits of agreement were about ±6 min^{-1}.

Alemdar and coauthors developed a smart textile based on macrobending hetero-core fiber optic for respiratory movement analysis [69,70]. They embedded within a textile, a periodic macrobending hetero-core fiber optic that formed a periodic "U". They experimentally assessed the increase of sensitivity to strain with the number of loops (ranging from one to seven loops) and the influence of the loop length on sensitivity. Then, the most sensitive configuration was tested to monitor the abdominal movement on a healthy volunteer.

Lau *et al.* designed a simple microbend fiber optic sensor for respiratory monitoring during MRI procedures [71]. They assessed the sensor feasibility during MRI procedures on twenty healthy volunteers, showing an accuracy of ±2 breaths per minute in the measurement of frequency rate. They also developed a system to monitor both respiratory rate and heart rate with a similar approach [72]. They tested the system on 11 volunteers during an MRI procedure, showing an accuracy of two breaths per minute for respiratory rate monitoring, and two beats per minute for heart rate monitoring.

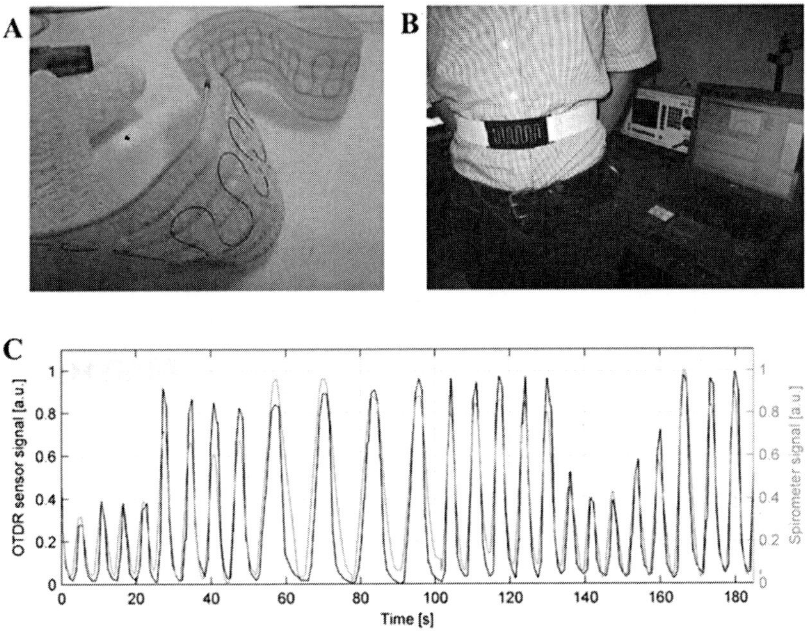

Figure 5. (**A**) Textiles based on macrobending FOS embedding a standard single mode fiber (adapted from [57]); (**B**) Macrobending sensors for monitoring of abdominal movements (adapted from [57]); (**C**) Output of the OTDR sensors during the monitoring of abdominal movement (adapted from [57]).

Recently, a smart textile based on the use of a periodic fiber optic microbend sensor has been proposed for heartbeat and respiratory monitoring [73]. The authors integrated a section of multimode optical fiber sandwiched between parallel strips acting as a microbender onto an elastic substrate. Basically, the system monitors respiration and heartbeat by detecting the vibration caused by these actions. The microbending results in light loss, which is detected by a photodetector. After a set ofin *vitro* experiments, they tested the textile on healthy volunteers. The measured heartbeat was compared to the one measured by a commercial device, the measured respiratory rate with the number of cycles manually counted in one minute.

The main characteristics of the abovementioned smart textiles and their applications are reported in Table 1.

4. DISCUSSION

Smart textiles are used in a number of industrial fields such as civil engineering, transport and medicine, amongst others. Regarding the application in medicine and healthcare, they allow the continuous monitoring of physiological parameters of great importance (e.g., respiratory activity and heartbeat).

The monitoring of these parameters can be performed by several approaches, and different devices are commercially available. The main advantage to integrate FOSs into smart textiles is the possibility to develop MR-compatible systems. The number of MR scanners, the request for high field devices and for MR procedures in several medical branches (e.g., cardiology, surgery, orthopedics and neurology) are increasing worldwide. This trend reflects the growing need for MR-compatible systems able to monitor physical parameters inside the scanner, in order to provide real-time feedback about the status of the patient.

As a consequence, smart textiles based on the use of FOSs can be considered a potential new market niche in the field of healthcare monitoring. In this scenario is important to have a small size and lightweight system, and more in general a system allowing a normal range of motion and preserve the patient's comfort. As a consequence, intensity-based fiber optic sensors seem to be indicated more than FBG ones that require the use of the interrogator. Moreover, also in homecare application the privacy and confidentiality of the patient must be respected. Solutions, such as enclosed rooms without traffic or others present and data transmission over secure lines can help to fulfill this requirement. Lastly, the problems of washability and, in some cases, sterilizability of FOS-based smart textiles are not well addressed. This feature must be considered because it can be a cost-effective option and can avoid the use of these systems for mono-patient or disposable applications. Therefore, future studies should address the robustness of these textiles against stress and washing cycles or their sterilizability The advantage is that FOSs are intrinsically safer than conventional sensors as they do not have any electrical connection with the patient and may be used at medium-term for monitoring different physiological parameters in order to replace standard sensors during MRI procedures. In this scenario it is also important to take into account the issue related to the connection of fibers. As reported by Kinet *et al.* [74] different solutions can be employed (e.g., commercially available standard connectors for optical fibers and *ad hoc* designed connectors). For the applications in the MRI environment, it is crucial to employ *ad hoc* designed, MR-compatible connectors.

Moreover, they can answer to the increasing demand for wearable systems for continuous monitoring of physiological parameters. A comfortable and wearable measuring system may significantly improve the quality of life of patients.

The main concerns related to the extensive use of smart textiles based on FOSs and their spread in commercially available products are related to the necessity to improve their stability due to light coupling (in particular in homecare long-term applications), and the abovementioned issue related to the connection of fibers (in particular in application during MRI procedures). The continuous improvement of the quality of the key optical components and their cost reduction, as well as the good metrological properties of these sensors motivates the increasing interest in this topic.

A recent review published by Quandt *et al.* in 2014 [18], described body-monitoring systems based on optical fibers and their application in the field of physiological monitoring. In this review, we devoted our attention especially to smart textiles based on FOSs for monitoring mechanical parameters of physiological interest. Also briefly reviewed are the working principles of the FOSs most used in these applications. Lastly, we focused on the metrological properties and on the performances of the systems reported in literature. In particular the most employed FOSs are based on FBG technology and on the modulation of intensity (in particular on macrobending). They are largely used for heartbeat and respiratory monitoring. These two kinds of sensors show complementary advantages and disadvantages in terms of sensitivity to movement and costs. FBG sensors are able to measure small strains thanks to their high sensitivity; on the other hand, in order to have good performances, interrogation must take place through an expensive device. The interrogation of intensity based FOSs is very simple and involves low cost and compact components, and their integration into textile fabrics may be very straightforward; on the other hand they are less sensitive than FBG sensors and are usually employed to monitor large movements (e.g., abdominal movements).

The use of smart textiles based on FOS in respiratory and heartbeat monitoring is just in its beginning stages, but their good metrological performances and the possibility to monitor patients in real time during MRI procedures motivates to continue research effort devoted to this topic.

AUTHOR CONTRIBUTIONS

Carlo Massaroni, Paola Saccomandi and Emiliano Schena contributed equally to this manuscript; Carlo Massaroni, Paola Saccomandi and Emiliano Schena wrote the article.

CONFLICTS OF INTEREST

The authors declare no conflict of interest.

REFERENCES

1. Lymberis, A.; Olsson, S. Intelligent Biomedical Clothing for Personal Health and Disease Management: State of the Art and Future Vision. *Telemed. J. E-Health* **2003**, *9*, 379–386.

2. Lobodzinski, S.S.; Laks, M.M. New devices for very long-term ECG monitoring. *Cardiol. J.* **2012**, *19*, 210–214.

3. Lobodzinski, S.S. ECG patch monitors for assessment of cardiac rhythm abnormalities. *Prog. Cardiovas. Dis.* **2013**, *56*, 224–229.

4. Inan, O.T.; Etemadi, M.; Wiard, R.M.; Giovangrandi, L.; Kovacs, G.T. Robust ballistocardiogram acquisition for home monitoring. *Physiol. Meas.* **2009**, *30*, 169–185.

5. Zhang, X.; Tao, X. Smart textiles: Passive smart. *Text. Asia* **2001**, 45–49.

6. Zhang, X.; Tao, X. Smart textiles: Active smart. *Text. Asia* **2001**, 49–52.

7. Krehel, M.; Wolf, M.; Boesel, L.F.; Rossi, R.M.; Bona, G.L.; Scherer, L.G. Development of a luminous textile for reflective pulse oximetry measurements. *Biomed. Opt. Express* **2014**, *5*, 2537–2547.

8. Li, H.N.; Li, D.S.; Song, G.B. Recent applications of fiber optic sensors to health monitoring in civil engineering. *Eng. Struct.* **2014**, *26*, 1647–1657.

9. Da Silva, A.F.; Gonçalves, A.F.; de Almeida Ferreira, L.A.; Araújo, F.M.M.; Mendes, P.M.; Correia, J.H. PVC smart sensing foil for advanced strain measurements. *IEEE Sens. J.* **2010**, *10*, 1149–1155.

10. Silvestri, S.; Schena, E. Optical-fiber measurement systems for medical applications. In *Optelectronics: Devices and Applications*; Predeep, P., Ed.; InTech: Rijeka, Croatia, 2011.

11. Lee, B. Review of the present status of optical fiber sensors. *Opt. Fiber Technol.* **2003**, *9*, 57–79.

12. Marazuela, M.; Moreno-Bondi, M. Fiber-optic biosensors—An overview. *Anal. Bioanal. Chem.* **2002**, *372*, 664–682.

13. Grattan, K.T.V.; Meggitt, B.T. *Optical Fiber Sensor Technology, Vol. 2—Devices and Technology*; Chapman & Hall: London, USA, 1998.

14. Udd, E. *Fiber Optic Sensors*, 1st ed.; John Wiley & Sons, Inc.: Hoboken, New Jersey, 2006.

15. Wang, X.D.; Wolfbeis, O.S. Fiber-optic chemical sensors and biosensors (2008–2012). *Anal. Chem.* **2012**, *85*, 487–508.

16. Krebber, K.; Liehr, S.; Witt, J. Smart technical textiles based on fibre optic sensors. In Proceedings of the 22nd International Conference on Optical Fiber Sensors, Beijing, China, 14 October 2012.

17. Ghosh, S.; Amidei, C.; Furrow, K. Development of a sensor-embedded flexible textile structure for apparel or large area applications. *Indian J. Fibre Text.* **2005**, *30*, 42–48.

18. Quandt, B.M.; Scherer, L.J.; Boesel, L.F.; Wolf, M.; Bona, G.L.; Rossi, R.M. Body-monitoring and health supervision by means of optical fiber-based sensing systems in medical textiles. *Adv. Healthc. Mater.* **2015**, *4*, 330–355.

19. Hill, K.O.; Fujii, Y.; Johnson, D.C.; Kawasaki, B.S. Photosensitivity in optical fiber waveguides: Application to reflection filter fabrication. *Appl. Phys. Lett.* **1978**, *32*, 647–649.

20. Meltz, G.; Morey, W.W.; Glenn, W.H. Formation of Bragg gratings in optical fibers by a transverse holographic method. *Opt. Lett.* **1989**, *14*, 823–825.

21. Rao, Y.J.; Webb, D.J.; Jackson, D.A.; Zhang, L.; Bennion, I. Optical in-fiber bragg grating sensor systems for medical applications. *J. Biomed. Opt.* **1998**, *3*, 38–44.

22. Gurkan, D.; Starodubov, D.; Yuan, X. Monitoring of the heartbeat sounds using an optical fiber bragg grating sensor. In Proceedings of the 4th IEEE Conference Sensors 2005, Irvine, CA, USA, 30 October–3 November 2005.

23. Haines, D.E. Determinants of lesion size during radiofrequency catheter ablation: The role of electrode-tissue contact pressure and duration of energy delivery. *J. Cariovasc. Electrophysiol.* **1991**, *2*, 509–515.

24. Yokoyama, K.; Nakagawa, H.; Shah, D.C.; Lambert, H.; Leo, G.; Aeby, N.; Ikeda, A.; Pitha, J.V.; Sharma, T.; Lazzara, R.;*et al*. Novel contact force sensor incorporated in irrigated radiofrequency ablation catheter predicts lesion size and incidence of steam pop and thrombus. *Circ. Arrhythm. Electrophysiol.* **2008**, *1*, 354–362.

25. Hao, J.Z.; Tan, K.M.; Tjin, S.C.; Liaw, C.Y.; Roy Chaudhuri, P.; Cuo, X.; Lu, Cu. Design of a foot-pressure monitoring transducer for diabetic patients based on FBG sensors. In Proceedings of the LEOS, the 16th Annual Meeting of the IEEE, Tucson, AZ, USA, 27–30 October 2003; pp. 23–24.

26. Al-Fakih, E.; Abu Osman, N.A.; Mahamd Adikan, F.R. The use of fiber Bragg grating sensors in biomechanics and rehabilitation applications: The state-of-the-art and ongoing research topics. *Sensors* **2012**, *12*, 12890–12926.

27. Tjin, S.C.; Tan, Y.K.; Yow, M.; Lam, Y.Z.; Hao, J. Recording compliance of dental splint use in obstructive sleep apnoea patients by force and temperature modeling. *Med. Biol. Eng. Comput.* **2001**, *39*, 182–184.

28. Saccomandi, P.; Schena, E.; Caponero, M.A.; di Matteo, F.M.; Martino, M.; Pandolfi, M.; Silvestri, S. Theoretical analysis and experimental evaluation of laser-induced interstitial thermotherapy in *ex vivo* porcine pancreas. *IEEE Trans. Biomed. Eng.* **2012**, *59*, 2958–2964.

29. Di Matteo, F.; Martino, M.; Rea, R.; Pandolfi, M.; Panzera, F.; Stigliano, E.; Schena, E.; Saccomandi, P.; Silvestri, S.; Pacella, C.M.; *et al*. US-guided application of Nd: YAG laser in porcine pancreatic tissue: An *ex vivo* study and numerical simulation. *Gastrointest. Endosc.* **2013**, *78*, 750–755.

30. Saccomandi, P.; Schena, E.; Giurazza, F.; del Vescovo, R.; Caponero, M.A.; Mortato, L.; Panzera, F.; Cazzato, R.L.; Grasso, F.R.; di Matteo, F.M.; *et al*. Temperature monitoring and lesion volume estimation during double-applicator laser-induced thermotherapy in *ex vivo* swine pancreas: A preliminary study. *Laser Med. Sci.* **2014**, *29*, 607–614.

31. Saccomandi, P.; Schena, E.; di Matteo, F.M.; Pandolfi, M.; Martino, M.; Rea, R.; Silvestri, S. Laser Interstitial Thermotherapy for pancreatic tumor ablation: Theoretical model and experimental validation. In

Proceedings of the 2011 Annual International Conference of the IEEE on Engineering in Medicine and Biology Society, EMBC, Boston, MA, USA, 30 August–3 September 2011; pp. 5585–5588.

32. Wehrle, G.; Nohama, P.; Kalinowski, H.J.; Torres, P.I.; Valente, L.C.G. A fibre optic Bragg grating strain sensor for monitoring ventilatory movements. *Meas. Sci. Technol.* **2001**, *12*, 805.

33. Allsop, T.; Miller, M.; Bennion, I.; Carroll, K.; Lloyd, G.; Webb, D.J. Application of long-period-grating sensors to respiratory plethysmography. *J. Biomed. Opt.* **2007**, *12*.

34. Cowie, B.M.; Webb, D.J.; Tam, B.; Slack, P.; Brett, P.N. Fibre Bragg grating sensors for distributive tactile sensing. *Meas. Sci. Technol.* **2007**, *18*.

35. Saccomandi, P.; Schena, E.; Oddo, C.M.; Zollo, L.; Silvestri, S.; Guglielmelli, E. Microfabricated Tactile Sensors for Biomedical Applications: A Review. *Biosensors* **2014**, *4*, 422–448.

36. Mishra, V.; Singh, N.; Tiwari, U.; Kapur, P. Fiber grating sensors in medicine: Current and emerging applications. *Sens. Actuat A Phys.* **2011**, *167*, 279–290.

37. Taffoni, F.; Formica, D.; Saccomandi, P.; Pino, G.D.; Schena, E. Optical fiber-based MR-compatible sensors for medical applications: An overview. *Sensors* **2013**, *13*, 14105–14120.

38. Watanabe, K.; Tajima, K.; Kubota, Y. Macrobending characteristics of a hetero-core splice fiber optic sensor for displacement and liquid detection. *IEICE Trans. Electron.* **2000**, *83*, 309–314.

39. Nishiyama, M.; Sasaki, H.; Watanabe, K. A deformation sensitive pad-structure embedded with hetero-core optic fiber sensors. *Sens. Actuat A Phys.* **2007**, *136*, 205–211.

40. Lekholm, A.; Lindstrom, L. Optoelectronic transducer for intravascular measurements of pressure variations. *Med. Biol. Eng.* **1969**, *7*, 333–335.

41. Wald, A.; Post, K.; Ransohoff, J.; Hass, W.; Epstein, F. A new technique for monitoring epidural intracranial pressure.*Med. Instrum.* **1977**, *11*, 352–354.

42. Piper, I.; Barnes, A.; Smith, D.; Dunn, L. The Camino intracranial pressure sensor: Is it optimal technology? An internal audit with a

review of current intracranial pressure monitoring technologies. *Neurosurgery* **2001**, *49*, 1158–1165.

43. Wolthuis, R.A.; Mitchell, G.L.; Hartl, J.C.; Saaski, E. Development of a dual function sensor system for measuring pressure and temperature at the tip of a single optical fiber. *IEEE Trans. Biomed. Eng.* **1993**, *40*, 298–302.

44. Babchenko, A.; Khanokh, B.; Shomer, Y.; Nitzan, M. Fiber optic sensor for the measurement of respiratory chest circumference changes. *J. Biomed. Opt.* **1999**, *4*, 224–229.

45. Maletras, F.X.; Augousti, A.T.; Mason, J. Construction and calibration of a new design of Fiber Optic Respiratory Plethysmograph (FORP). In Proceedings of the SPIE 4444, Optomechanical Design and Engineering 2001, San Diego, CA, USA, 29 July 2001; p. 285.

46. Nishyama, M.; Miyamoto, M.; Watanabe, K. Respiration and body movement analysis during sleep in bed using hetero-core fiber optic pressure sensors without constraint to human activity. *J. Biomed. Opt.* **2011**, *16*.

47. Schena, E.; Saccomandi, P.; Silvestri, S. A high sensitivity fiber optic macro-bend based gas flow rate transducer for low flow rates: Theory, working principle, and static calibration. *Rev. Sci. Instrum.* **2013**, *84*.

48. Saccomandi, P.; Schena, E.; Silvestri, S. A novel target-type low pressure drop bidirectional optoelectronic air flow sensor for infant artificial ventilation: Measurement principle and static calibration. *Rev. Sci. Instrum.* **2011**, *82*.

49. Husdi, I.R.; Nakamura, K.; Ueha, S. Sensing characteristics of plastic optical fibres measured by optical time-domain refelectometry. *Meas. Sci. Technol.* **2004**, *15*, 1553–1559.

50. Augousti, A.T.; Maletras, F.X.; Mason, J. Improved fibre optic respiratory monitoring using a figure-of-eight coil.*Physiol. Meas.* **2005**, *26*, 585–590.

51. Augousti, A.T.; Raza, A.; Graves, M. Design and characterisation of a fibre optic respiratory plethysmograph (FORP). In Proceedings of the SPIE 1996, Biomedical Sensing, Imaging, and Tracking Technologies I, San Jose, CA, USA, 27 January 1996; pp. 250–257.

52. Narbonneau, F.; Kinet, D.; Paquet, B.; Depré, A.; de Jonckheere, J.; Logier, R.; Zinke, J.; Witt, J.; Krebber, K. Smart textile embedding

optical fibre sensors for healthcare monitoring during MRI. *Adv. Sci. Tech.* **2008**, *60*, 134–143.

53. Grillet, A.; Kinet, D.; Witt, J.; Schukar, M.; Krebber, K.; Pirotte, F.; Depré, A. Optical fiber sensors embedded into medical textiles for healthcare monitoring. *IEEE Sens. J.* **2008**, *8*, 1215–1222.

54. De Jonckheere, J.; Jeanne, M.; Narbonneau, F.; Witt, J.; Paquet, B.; Kinet, D.; Kreber, K.; Logier, R. OFSETH: A breathing motions monitoring system for patients under MRI. In Proceedings of the Annual International Conference of the IEEE Engineering in Medicine and Biology Society, Minneapolis, USA, 2-6 September 2009; pp. 1016–1019.

55. Narbonneau, F.; D'Angelo, L.T.; Witt, J.; Paquet, B.; Kinet, D.; Kreber, K.; Logier, R. FBG-based smart textiles for continuous monitoring of respiratory movements for healthcare applications. In Proceedings of the 12th IEEE International Conference on e-Health Networking Applications and Services (Healthcom), Lyon, France, 1–3 July 2010; pp. 277–282.

56. Witt, J.; Narbonneau, F.; Schukar, M.; Krebber, K.; de Jonckheere, J.; Jeanne, M.; Kinet, D.; Paquet, B.; Depré, A.; D'Angelo, L.T.; *et al.* Medical textiles with embedded fiber optic sensors for monitoring of respiratory movement.*IEEE Sens. J.* **2012**, *12*, 246–254.

57. Krebber, K. Smart Technical Textiles Based on Fiber Optic Sensors. In *Current Developments in Optical Fiber Technology*; InTech: Rijeka, Croatia, 2013; pp. 319–344.

58. Silva, A.F.; Carmo, J.P.; Mendes, P.M.; Correia, J.H. Simultaneous cardiac and respiratory frequency measurement based on a single fiber Bragg grating sensor. *Meas. Sci. Tech.* **2011**, *22*.

59. Carmo, J.P.; da Silva, A.M. F.; Rocha, R.P.; Correia, J.H. Application of fiber Bragg gratings to wearable garments. *IEEE Sens. J.* **2012**, *12*, 261–266.

60. Dziuda, Ĺ.; Skibniewski, F.W.; Krej, M.; Baran, P.M. Fiber Bragg grating-based sensor for monitoring respiration and heart activity during magnetic resonance imaging examinations. *J. Biomed. Opt.* **2013**, *18*.

61. Dziuda, Ĺ.; Skibniewski, F.W.; Krej, M.; Lewandowski, J. Monitoring respiration and cardiac activity using fiber Bragg grating-based sensor. *IEEE Trans. Biomed. Eng.* **2012**, *59*, 1934–1942.

62. Dziuda, L.; Krej, M.; Skibniewski, F.W. Fiber Bragg grating strain sensor incorporated to monitor patient vital signs during MRI. *IEEE Sens. J.* **2013**, *13*, 4986–4991.

63. Krej, M.; Dziuda, L.; Skibniewski, F. A method of detecting heartbeat locations in the ballistocardiographic signal from the fiber-optic vital signs sensor. *IEEE J. Biomed. Health Inform.* **2015**.

64. Allsop, T.; Bhamber, R.; Lloyd, G.; Miller, M.R.; Dixon, A.; Webb, D.; Castañón, J.D. A.; Bennion, I. Respiratory function monitoring using a real-time three-dimensional fiber-optic shaping sensing scheme based upon fiber Bragg gratings. *J. Biomed. Opt.* **2012**, *17*.

65. Li, H.; Yang, H.; Li, E.; Liu, Z.; Wei, K. Wearable sensors in intelligent clothing for measuring human body temperature based on optical fiber Bragg grating. *Opt. Express* **2012**, *20*, 11740–11752.

66. Yoo, W.J.; Jang, K.W.; Seo, J.K.; Heo, J.Y.; Moon, J.S.; Park, J.Y.; Lee, B.S. Development of respiration sensors using plastic optical fiber for respiratory monitoring inside MRI system. *J. Opt. Soc. Korea* **2010**, *14*, 235–239.

67. Krehel, M.; Rossi, R.M.; Bona, G.L.; Scherer, L.J. Characterization of flexible copolymer optical fibers for force sensing applications. *Sensors* **2013**, *13*, 11956–11968.

68. Krehel, M.; Schmid, M.; Rossi, R.M.; Boesel, L.F.; Bona, G.L.; Scherer, L.J. An optical fibre-based sensor for respiratory monitoring. *Sensors* **2014**, *14*, 13088–13101.

69. Alemdar, K.; Likoglu, S.; Fidanboylu, K.; Toker, O. A novel periodic macrobending hetero-core fiber optic sensor embedded in textiles. In Proceedings of the SPIE 9062, Smart Sensor Phenomena, Technology, Networks, and Systems Integration, San Diego, CA, USA, 8 March 2014.

70. Alemdar, K.; Likoglu, S.; Fidanboylu, K.; Toker, O. A novel periodic macrobending hetero-core fiber optic sensor embedded in textiles. In Proceedings of the 8th International Conference on Electrical and Electronics Engineering (ELECO), Bursa, Turkey, 28–30 November 2013; pp. 467–471.

71. Lau, D.; Chen, Z.; Teo, J.T.; Ng, S.H.; Rumpel, H.; Lian, Y.; Yang, H.; Kei, P.L. Intensity-Modulated Microbend Fiber Optic Sensor for Respiratory Monitoring and Gating During MRI. *IEEE Trans. Biomed. Eng.* **2013**, *60*, 2655–2662.

72. Chen, Z.; Lau, D.; Teo, J.T.; Ng, S.H.; Yang, X.; Kei, P.L. Simultaneous measurement of breathing rate and heart rate using a microbend multimode fiber optic sensor. *J. Biomed. Opt.* **2014**, *19*.

73. Yang, X.; Chen, Z.; Chia, S.; Lam, H.; Ng, S.; Teo, J.; Wu, R. Textile Fiber Optic Micro Bend Sensor Used for Heartbeat and Respiration Monitoring. *IEEE Sens. J.* **2015**, *15*, 757–761.

74. Kinet, D.; Mégret, P.; Goossen, K.W.; Qiu, L.; Heider, D.; Caucheteur, C. Fiber Bragg Grating Sensors toward Structural Health Monitoring in Composite Materials: Challenges and Solutions. *Sensors* **2014**, *14*, 7394–7419.

CHAPTER 4

Medical Textiles as Vascular Implants and Their Success to Mimic Natural Arteries

Charanpreet Singh [1,†], Cynthia S. Wong [1,†] and Xungai Wang [1,2,†,*]

[1]Australian Future Fibres Research and Innovation Centre, Institute for Frontier Materials, Deakin University, Geelong, VIC 3216, Australia

[2]School of Textile Science and Engineering, Wuhan Textile University, Wuhan 430073, China

ABSTRACT

Vascular implants belong to a specialised class of medical textiles. The basic purpose of a vascular implant (graft and stent) is to act as an artificial conduit or substitute for a diseased artery. However, the long-term healing function depends on its ability to mimic the mechanical and biological behaviour of the artery. This requires a thorough understanding of the structure and function of an artery, which can then be translated into a synthetic structure based on the capabilities of the manufacturing method utilised. Common textile manufacturing techniques, such as weaving, knitting, braiding, and electrospinning, are frequently used to design vascular implants for research and commercial purposes for the past decades. However, the ability to match attributes of a vascular substitute to those of a native artery still remains a challenge. The synthetic implants have been found to cause disturbance in biological, biomechanical, and hemodynamic parameters at the implant site, which has been widely attributed to their structural design. In this work, we reviewed the design aspect of textile vascular implants and compared them to the structure of a natural artery as a basis for assessing the level of success as an implant. The outcome of this work is expected to encourage future design strategies for developing improved long lasting vascular implants.

KEYWORDS

vascular stent; graft; artery; weaving; knitting; electrospinning; braiding; compliance; non-linearity; anisotropy

1. INTRODUCTION

The use of textiles for medical applications can be traced back to the early ages in wound care applications such as sutures and wound dressings. The suitability of textiles, whether in fibre or fabric form, lies in their structural flexibility, whereby some exhibit properties similar to human tissues, which are also composed of fibrous components. Another advantage of textile-based substrates is their design flexibility (from fibre to fabric stage), which can be modified to emulate the mechanical behaviour (elasticity, strength, stiffness, fluid permeability) of native biological tissue. These properties give textiles an edge over other materials (metals and plastics) in the area of soft tissue repair, for example, cardiovascular implants, which are used to replace/repair diseased arteries.

Cardiovascular implant market (grafts and stents) is growing at a fast pace due to increasing number of patients with vascular diseases and limited biological replacement options available. Thus, synthetic implants offer off-the-shelf solution in a range of design specifications. During the last two decades, a significant amount of research and industry effort has been put into developing vascular implants intended for various anatomical locations. Among the soft tissue implants, design optimisation of vascular implants is considered as one of the most complex tasks and has therefore been a continuing challenge for biomedical device engineers. The reason for this is the inability to match biomechanical behaviour of a synthetic implant to that of an artery. This is due to the heterogeneous structure of an artery, which imparts unique mechanical features (non-linearity, anisotropy, viscoelasticity, compliance) in the vascular wall. An artery shows very low stress response at low pressures and exhibits a steep increase in elastic modulus as the pressure is increased, a property known as non-linearity or incremental elastic modulus (Figure 1). Simultaneously, geometrical arrangement of structural components in the vessel wall imparts anisotropy characteristic to its mechanical behaviour. On the other hand, a synthetic arterial substitute (woven Dacron® graft) is a homogenous structure, which exhibits low elasticity and relatively higher linear stress response at similar levels of strain (Figure 1) [1]. The viscoelasticity property is another important component that determines the hemodynamic behaviour of an arterial vessel and is an equally difficult feature to mimic in a synthetic implant. When an artery is subjected to cyclic inflation-deflation stresses, it does not react instantaneously to these stresses [2,3]. Instead, the artery, due to it being viscoelastic, produces a delayed response known as hysteresis when subjected to changes in pressure and volume (Figure 1). In addition to the biomechanical properties mentioned above, compliance, defined as radial extensibility of an artery in the physiological pressure range (80–120 mmHg) plays a decisive role in vascular mechanics. In a synthetic implant, this property is directly related to structural construction and material property, and hence highly variable. An unmatched compliance between an artery and implant is a common cause of long-term complications which result in ultimate failure of the surgical procedure due to altered pressure and flow dynamics [4,5]. Therefore, the design feature of medical textiles is critical when they are being considered for vascular implant applications.

Synthetic vascular implants are currently manufactured using standard textile manufacturing techniques such as weaving, knitting, braiding and electrospinning. A number of reviews have reported on the mechanical property comparisons of different types of vascular grafts and their clinical performance [6,7,8]. In 1986, Pourdeyhimi and Wagner presented an extensive review focusing on structures of the synthetic grafts to explain the reported clinical observations of these grafts [9,10]. However, recent studies on grafts and stents rarely consider the effect manufacturing techniques have on the structure. This paper will discuss the structure of vascular implants produced by different textile manufacturing methods and analyse them with respect to the arterial wall structure. The analysis is done while following the development timeline of these manufacturing techniques in the vascular implant industry. This approach will provide an understanding of the degree of success, which has been achieved by textile-based implants in mimicking the mechanics of native arteries since their first clinical introduction. Some innovative design concepts, which have attempted to reduce the mechanical property mismatch between the implant and the host artery, are also discussed. These works are thought to encourage the design of longer lasting vascular implants in future.

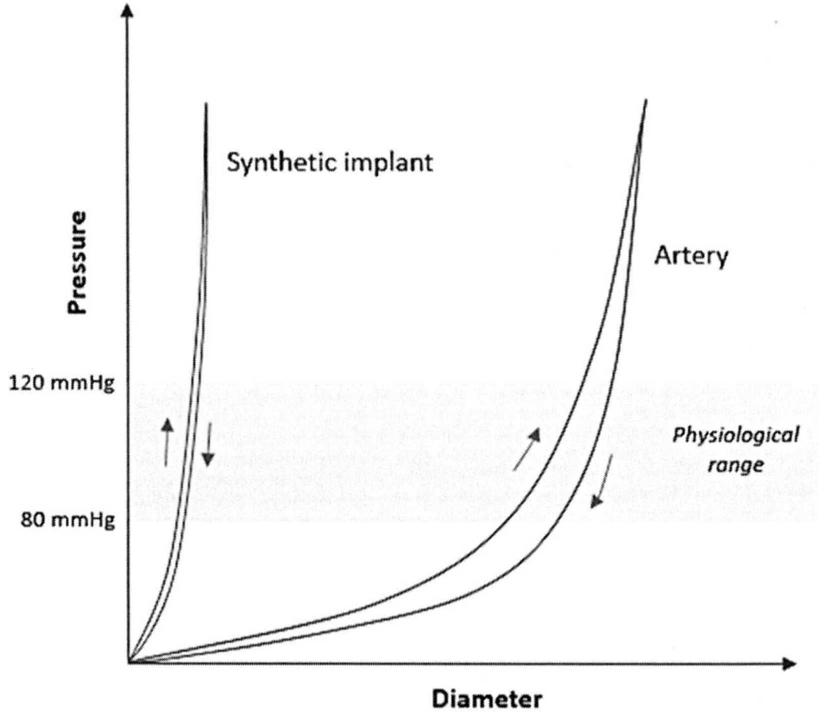

Figure 1. Comparison of pressure-diameter curves between an artery and a synthetic implant.

2. ARTERY: STRUCTURE AND MECHANICAL BEHAVIOUR

The primary step towards designing a vascular implant with improved mechanical response is to understand the arterial structure, its components and mechanical role of each component. An arterial wall is composed of three major layers namely, tunica intima, tunica media, and tunica adventitia [11]. The intima or the endothelial layer comprises of a single layer of endothelial cells. The tunica media is the thickest layer and composed of circumferentially arranged elastic fibres, smooth muscle cells, and collagen fibres. The outermost adventitia layer comprises of large diameter collagen fibres oriented longitudinally as wavy bundles. The mechanical behaviour of an artery is based mainly on the thickness of media layer and its main structural components (elastin and collagen), which differ significantly in their elastic modulus (elastin = 0.6–1 MPa, collagen = 1 GPa) [2,3]. The function of both these components was investigated by Roach and Burton by selectively dissolving them and comparing the mechanical behaviour of the altered artery [1]. The study demonstrated that the initial slope in a typical stress-strain curve of an artery was contributed entirely by elastin fibres, while the final slope was due to collagen fibre stress response (Figure 2). The mid region (upturning region) involves successive transition or shifting of load from elastin to collagen and corresponds to the normal *in vivo* operating range (80–120 mmHg) of an artery. The non-linearity or incremental elastic modulus property is due to the wavy and random configuration of elastin and collagen fibres when unpressurised. With increase in pressure, elastin and collagen fibres start to straighten progressively. Elastin fibres become nearly straight at lower end of physiological pressure (80 mmHg). A further increase in pressure results in stretching of elastin and successive straightening as well as stretching of collagen fibres until top end of the physiological range (120 mmHg) is reached. An increase in pressure beyond this region (upturning region) results in fully stretched collagen and elastin fibres where stress response of collagen fibres dominates the arterial behaviour. This non-linear stress-strain behaviour is considered to be the key to elastic stability in arteries, which protects them from developing pathological conditions of aneurysms and 'blowout' at high pressures. The mechanical behaviour of arteries is also controlled by relative proportion of collagen and elastin fibres. High elastin content in ascending aorta (41% of dry weight) compared to descending aorta (30% of dry weight) is the reason behind a decreasing compliance trend observed while moving downstream in an aortic vessel [12,13].

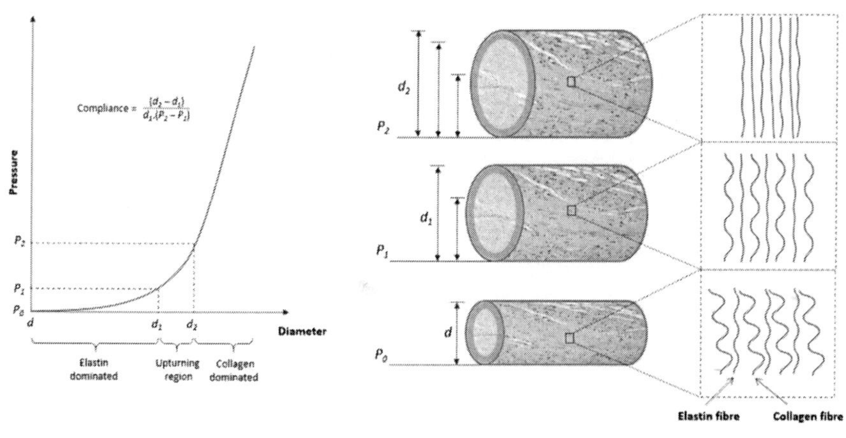

Figure 2. Role of fibrous components (elastin and collagen) in shaping the pressure-diameter relation of an artery.

The mechanical performance of an artery in the upturning region is expressed in terms of compliance value (percentage increase in diameter for a given increase in pressure) (Figure 2). A high compliance value indicates optimum level of structural performance for maintaining pulsatile blood flow. The property of an artery to support pulsatile flow is also known as the windkessel function, and is critically important in arteries in proximity of the heart such as the aorta [14]. During systole, the heart pumps nearly 60–100 mL of blood into the aorta while there is no supply during diastole. The windkessel function assists in temporarily storing a portion (approx. 50%) of systolic blood volume in aorta which can be later used during diastole to maintain consistent blood flow throughout the arterial network (Figure 3). During this process, the cross-sectional area of aorta (ascending aorta) can increase by 11% [15]. The basis of windkessel function lies in the compliance property, as a non-compliant (stiff) aortic vessel cannot expand sufficiently to store blood.

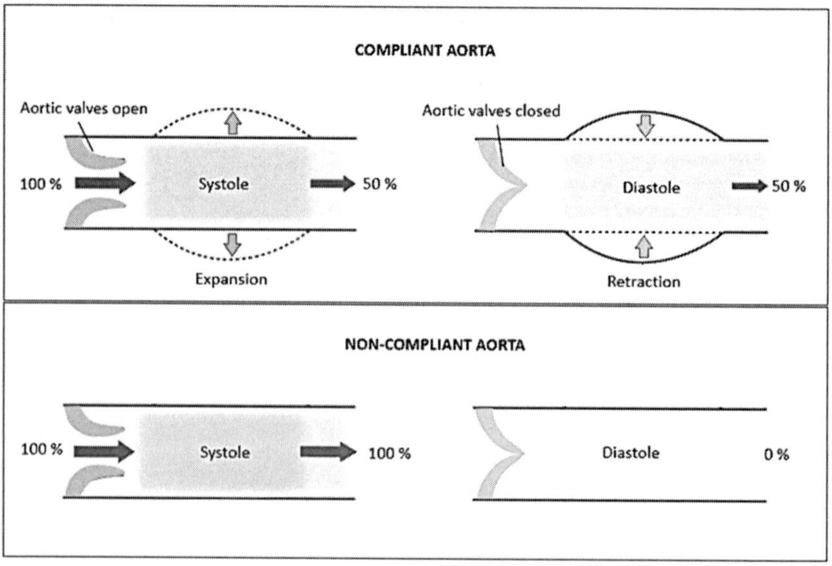

Figure 3. The role of compliance in the windkessel function of aorta.

Anisotropy is another important characteristic of arterial wall attributed to circumferential orientation of collagen fibres and their unequal orthogonality in unstressed state. Arterial tissue is 40% stiffer in circumferential compared to axial direction and approximately 100% stiffer when inflated to physiological pressures [16]. Viscoelasticity or stress relaxation is a decisive component of arterial elasticity property. When an artery is subjected to cyclic inflation-deflation stresses, viscoelasticity causes hysteresis in pressure-volume curve. The area enclosed by hysteresis loop represents the energy lost in each cycle, which indicates that a major component of strain energy is recovered elastically each time arterial wall is distended [2]. The dissipation of strain energy due to viscoelasticity assists in attenuating forward pressure pulses, which propagate along arteries as waves of circumferential distention. Additionally, viscoelastic property has been suggested to improve the fatigue life of arteries by reducing dynamic stresses and strains in the wall [17].

3. TEXTILE STRUCTURES AS VASCULAR IMPLANTS

The incorporation of textiles as vascular implants started in 1952 with the pioneering work of Voorhees and colleagues who replaced diseased aortic vessels of dogs with woven Vinyon-N (a polyvinylchloride) tubes [18,19]. Within few years of Voorhees's work, a number of studies reported clinical trials with different types of materials (Nylon, Teflon®, Dacron®, Orlon®), and constructions (woven, knitted, braided) in various diameters (6–20 mm) [20,21]. The role of manufacturing technique should be considered when deciding on parameters such as the handling and *in vivo* behaviour of the grafts. The definition of an ideal vascular implant requires it to be (1) biocompatible, (2) non-thrombogenic, (3) compliant, (4) fatigue resistant, (5) flexible yet robust, (6) readily available, and (7) easy to manufacture. Among these, attributes 3–7 are influenced by the manufacturing method used and structural design of the implant, while attributes 1–2 are contributed by material selection. There are a number of techniques used to manufacture medical textiles and those that are used for vascular implants are mainly weaving, knitting, braiding and electrospinning. Each of these techniques presents specific characteristics and benefits which led to varying suitability for different types of vascular implants.

3.1. Weaving

A woven graft is manufactured by interlacing two sets of yarn (warp and weft) oriented at 90° to each other. These grafts are currently available in different types of weave designs namely, plain, twill, and satin (Figure 4). The main characteristics which were considered important while introducing these grafts in clinical practice were surface smoothness, handling ease, non-reactivity, water permeability, bursting strength, suture retention strength, and biological healing response [22,23,24,25,26]. On the other hand, biomechanical optimisation of graft design was largely overlooked in the last six decades since the conception of this great innovation. This is also indicated by the fact that the basic construction (a homogenous single layer woven fabric) of a woven graft has remained the same since its introduction in the 1950s.

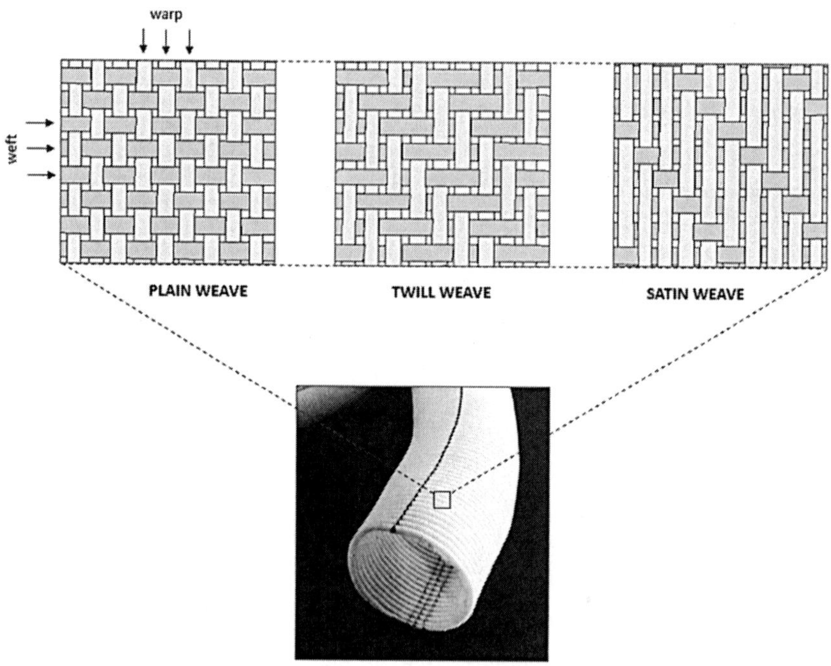

Figure 4. Structural design patterns of a woven Dacron® graft.

A commercial woven graft can be considered similar to a windcheater jacket from structural and material perspective, and bears no relationship to the composite design principle of an artery. The dissimilarity is apparent when the fibre itself is analysed, whereby the elastic behaviour of polyester fibres corresponds to collagen fibres while no graft component performs the role of elastin fibre, *i.e.*, low stress elastic response (Table 1). Furthermore, at the structural level, the straight yarn interlacement in woven grafts is unlike the helically arranged wavy configuration of collagen fibres, which induces non-linearity property in an artery. The orthogonal arrangement of both yarn components results in low axial elongation and poor radial compliance properties in these grafts [27]. The desired axial elasticity is often achieved through graft crimping, which also helps maintain tubular shape of graft during bending. However, optimum compliance is still an unachieved goal even with numerous proofs of its clinical importance [28,29,30].

Table 1. Elastic modulus of components of an artery [31] and a woven Dacron® graft. (* Stainless steel strut in a woven stent-graft).

Structure	Component	Elastic Modulus (MPa)
Artery	Elastin	0.6–1
	Collagen	1000
	Smooth Muscle	0.1
Woven graft	Dacron® Polyester	800–900
	Stainless steel*	190×10^3–210×10^3

The consequences of poor compliance can be fatal as it can change the transmission characteristics of pulse waves [4,32,33]. Moreover, a compliant host artery expands more than the stiff graft, which creates an abnormal stretch on the suture line. The continuous stretch subsequently produces structural fatigue of the arterial wall at the anastomosis [5]. A significant amount of research has been done on optimising construction and handling properties of woven grafts [34,35,36,37,38,39] but radial compliance/elasticity is still an unresolved issue [27,40] and has a direct influence on long term patency of these grafts (Table 2).

Table 2. Comparison of compliance property of Dacron® grafts with natural blood vessels. Adapted from [33].

Structure	Compliance (mmHg $\times 10^{-2}$)
Artery	7.4
Vein	2.7
Dacron® Polyester (Woven)	1.9
Dacron® Polyester (Knitted)	2.3

There are very few studies that report initiatives to match graft mechanics to the host artery. These include studies published soon after the work of

Voorhees, which attempted to tackle graft plasticity by using elastic filaments (Lycra®spandex) [41], and crimped synthetic filaments (Helanca® nylon and Dacron® polyester) [42,43]. Their concept was based on mimicking the role of wavy collagen and elastin fibres by using crimped and elastic yarns, respectively. Later, the combined use of both these components in plain weave constructions was successfully reported [44,45]. This approach improved the compliance nearly 17 times compared to a commercial woven graft (0.0324 ± 0.0083 kPa^{-1} *vs.* 0.00186 ± 0.0005 kPa^{-1}) and the new compliance matched that of a human common carotid artery (0.0238 ± 0.0132 kPa^{-1}) [44]. However, from the design perspective, woven graft was still in its primitive single layer fabric stage.

In 2011, an innovative woven graft design concept was reported by Chen *et al.* based on the coaxial graft concept of Sonoda *et al.* [46,47]. In this work, a bilayer woven graft prototype was developed in which the inner layer was constructed from low modulus yarns (poly-trimethylene terephthalate) while the outer layer made from high modulus yarns (polyester) was stitched in a crimped form to the inner layer (Figure 5). This type of structure can mimic the layered structure of blood vessels in a way that the strain at lower pressures is absorbed by the inner layer while a further increase in pressure will cause the outer layer to uncrimp and "join-in" with the inner layer to increase the overall elastic modulus of the graft. The authors reported a similar observation where elastic modulus of the new graft remained low, up to 80 mmHg and increased rapidly afterwards [46]. However, a drawback of this design is reduced bending flexibility or kink resistance due to the crimped outer layer, which may limit its clinical applicability.

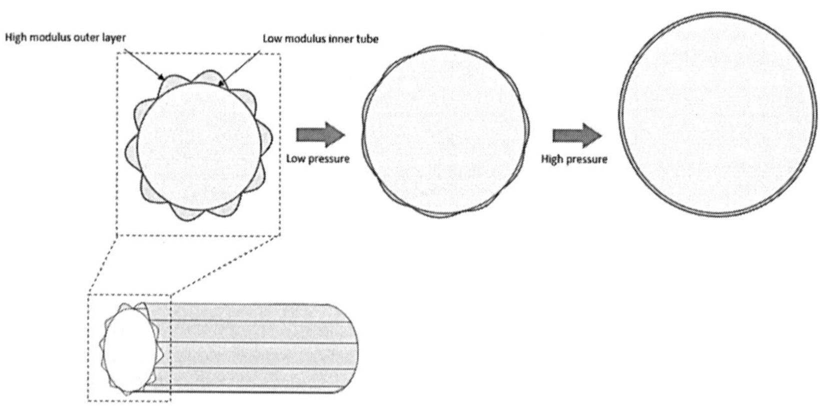

Figure 5. An explanation of thebilayer woven graft design concept proposed by Chen*et al.* [46].

While *in vitro* studies can provide proof of concept confirmation for new designs, *in vivo* studies, whether on animals or human beings, provide more convincing evidence of their performance in actual clinical environment. One of the early long-term animal studies established the advantage of bi-component graft design (spandex and Dacron®) in improving graft mechanical properties [45]. The graft was sufficiently compliant (6.9%, 80/100 mmHg; 9.5%, 80/200 mmHg) even after one year of implantation and did not show any plastic dilatation at explantation after three years. Unfortunately, no published *in vivo* trial data is available for rest of the new designs discussed above [44,46,47].

3.2. Knitting

Knitted grafts have a looped filament construction in which a continuous interconnecting chain of yarn loops spirals around the graft circumference. Knitted structures are softer, more flexible, compliant, and have better handling characteristics than woven structures. The most common types of knits, which are used for graft design, are the weft knit and warp knit constructions (Figure 6). Warp knitted structures have less stretch than weft knits, and therefore are inherently more dimensionally stable. The knitted grafts were first introduced in clinical practice in 1955 with the intent to remove the seam problem in a woven graft [48]. Later, in 1958, the renowned heart surgeon Dr. Michael DeBakey firmly established the clinical usefulness of knitted Dacron® grafts [49,50]. The trend of using Dacron® as a standard graft material started thereafter owing to the better long-term biostability of Dacron® compared to other available materials (Nylon and Orlon®). Improved stretch widthwise (radial compliance) compared to woven grafts was a promising observation in knitted grafts. However, high porosity and long-term dilatation were also reported [26,41]. The reason being absence of structural heterogeneity in knitted grafts as in arteries which protects them from undergoing fatigue dilation. Since then, several studies have been published reporting the use of different coating techniques, constituent materials, and knitted patterns with a motive to improve handling, fatigue and biological performance of these grafts [51,52,53,54,55,56,57,58,59]. Based on the inherent structural flexibility of knitted structures, many investigations have reported their use as elastic tubular substrates [60,61,62,63,64]. This property also formed the reason behind the use of monofilament knitted mesh structures as vascular stents [65,66,67,68,69]. However, commercially available knitted implants do not vary much in their structure, which is mainly a plain knit single layer homogenous structure bearing resemblance to an upholstery fabric rather than to an artery. In addition, Dacron® has been the only material for commercial knitted grafts since its use was first proposed by DeBakey [49].

Although Dacron® grafts are currently used in the clinics, there is a significant mismatch between their mechanical behaviour in comparison to natural arteries (Table 2) [40].

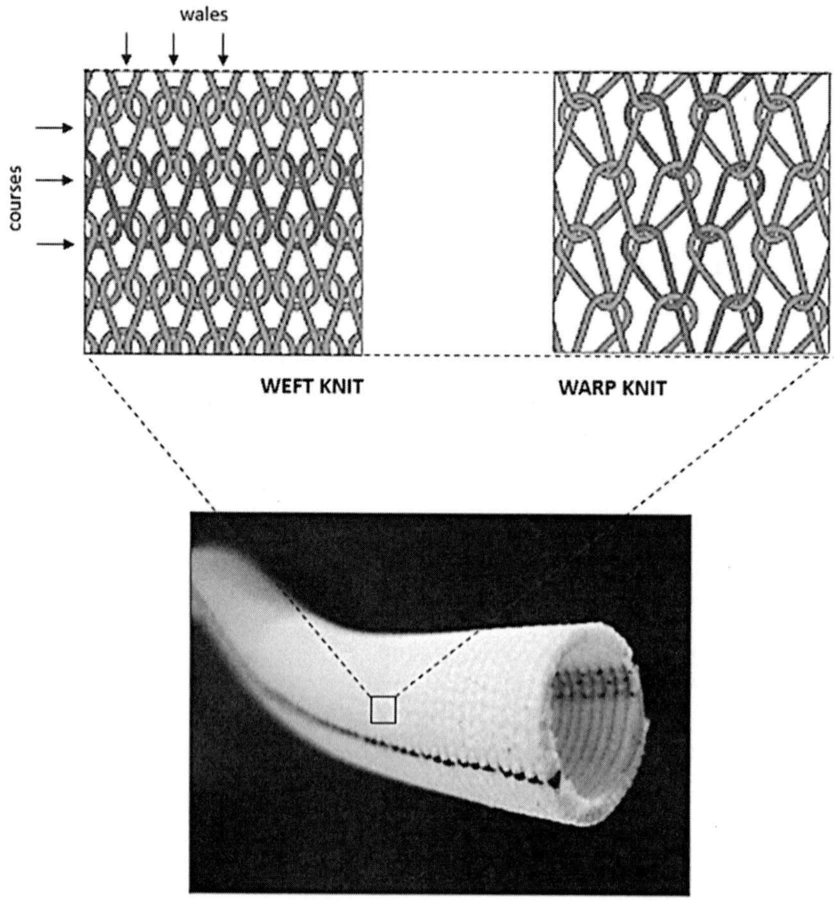

Figure 6. Structural design patterns of a knitted Dacron® graft.

Similar to woven grafts, knitted grafts also underwent design trials to improve their elastic behaviour. The first report on use of spandex filament knitted graft as a dog abdominal aorta (diameter 8–10 mm) was presented by Wagner *et al.* [41]. However, the solo use of spandex fibre as graft material was observed to cause long-term dilatation defect in the graft and is attributed to homogenous single layer structure of the graft. Some of the latest studies tried to use a composite polyester/spandex filament yarn to improve graft elasticity [62,64]. This type of material composition allows load sharing among both components and can prevent dilatation issues if structural design pattern is also modified. However, these studies only report basic improvements in mechanical properties of a plain weft knit structure and lack the ability to be considered as a significant design improvement to mimic arterial mechanics in a knitted graft.

In a recent study, an innovative knitted stent-graft design was reported which closely mimics the natural artery mechanical behaviour [70]. The design is based on the concept of longitudinal structural segmentation or metamerism in which the knitted tube is divided into multiple low and high modulus segments arranged in alternating sequence (Figure 7). The low modulus (knitted polyurethane) sections tend to remain contracted (reduced diameter) when unpressurised while high modulus (knitted polyester) maintain the as-knit configuration. Therefore, at low internal pressure, the expansion of low modulus segments controls the stress response of the knitted tube until their circumference equals that of high modulus segments. At high pressures, the combined response of both the segments increases the stress response sharply, exhibiting an incremental elastic modulus property similar to natural arteries. The low modulus segments act as intermittent "buffer zones" which assist in radial expansion as well as provide a kink-free configuration to the knitted tube. The compliance of this new design (volumetric: 0.056 ± 0.006 mL/mmHg; radial: 9.8×10^{-4} $mmHg^{-1}$) is nearly 7 and 15 times better compared to a conventional knitted stent (radial: 1.45×10^{-4} $mmHg^{-1}$) and a commercial woven Dacron® graft (volumetric: 0.0038 ± 0.002 mL/mmHg), respectively, and falls well within the physiological range of aortic vessel [70,71]. However, the *in vivo*performance of this design is still unavailable to demonstrate its clinical performance.

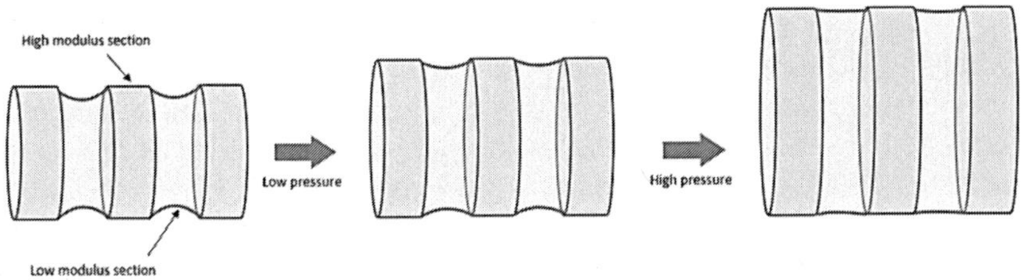

Figure 7. The segmented design concept as proposed by Singh and Wang to improve the compliance property of a knitted vascular implant [70,71].

3.3. Braiding

The application of braided textiles in traditional application areas (apparel, upholstery) is very limited compared to knitted and woven structures and is mainly found in technical textiles (industrial, sports, automotive). Braiding technique involves the use of three or more component yarns, which are intertwined at an angle to each other. In other words, braided structures are similar to traditional woven structures but with an angle bias. The angled mesh structure allows easy radial expansion of braided tubes. This property formed the basis of using braiding in the early years of textile graft innovation [72,73,74,75].

The advent of endovascular technology in the 1990s led to the search for user-friendly stent designs [76]. The ability of tubular braided structures to compress easily and recoil back made them a promising candidate for self-expanding vascular stents [77]. Low bending rigidity of braided mesh tube was also an additional advantage for stent application. However, radial expansion in braided tubes occurred at the expense of axial shortening and elastic recovery was limited to open mesh braids only. Also, braided stents were not very successful in endograft application as it required suturing/bonding of a tubular graft to stent wires and, hence, limited the movement of constituent wires at their cross-over points which ultimately led to stent failure (wire breakage) under cyclic, *in vivo* conditions [78]. This limitation shifted braided stents towards stent applications which do not require graft covering, suitable for peripheral atherosclerotic arteries, which is also their current main application area (Table 3) [79].

The commercially available braided stents are made from various metallic alloys (Table 3), which make them highly non-compliant and mechanically similar to a braided plumbing hose (Figure 8). Since low porosity is an important requirement in flow diverter stents, the number of wires is generally kept high (up to 96) to increase mesh density [80,81,82,83]. Similarly, compression of atherosclerotic region requires high radial compression strength, which is generally achieved by increasing mesh density. However, an inverse effect of high mesh density is increased stent stiffness (radial and longitudinal), which has a negative impact on the host artery hemodynamics [84,85]. The arterial length covered by the stent becomes non-compliant and results in pressure attenuation across this region and localised increase in pulse-wave velocity. A stiff metallic stent also causes straightening of the host artery and may induce kinks [86]. The development timeline of braided vascular stents to date mainly includes trials with various polymeric filaments [65,87,88,89,90,91,92,93,94,95,96] and stent parameter optimisation studies [89,96,97,98,99,100], while improvement in conventional braided design (a single component homogenous construction) to match arterial mechanics was rarely investigated.

Table 3. Categorisation of commercial braided stents according to their application area, material and design feature.

Trade Name	Manufacturer	Application	Material	Design Feature
PIPELINE®	ev3 Inc.	Flow diverter stent	Cobalt-Chromium + Platinum	Single layer braided tube
p64®	Phenox GmbH		Nitinol	
LEO® PLUS	Balt Extrusion		Nitinol	
SILK®			Nitinol + Platinum	
LVIS Device®	MicroVention Inc.		Nitinol	
WALLSTENT™	Boston Scientific Co.	Carotid stent	Elgiloy®	
ROADSAVER®	Terumo		Nitinol	
SUPERA®	Abbott Vascular	Peripheral stent	Nitinol	
Agili-D®	Altura Medical	Abdominal endograft	Metal alloy	

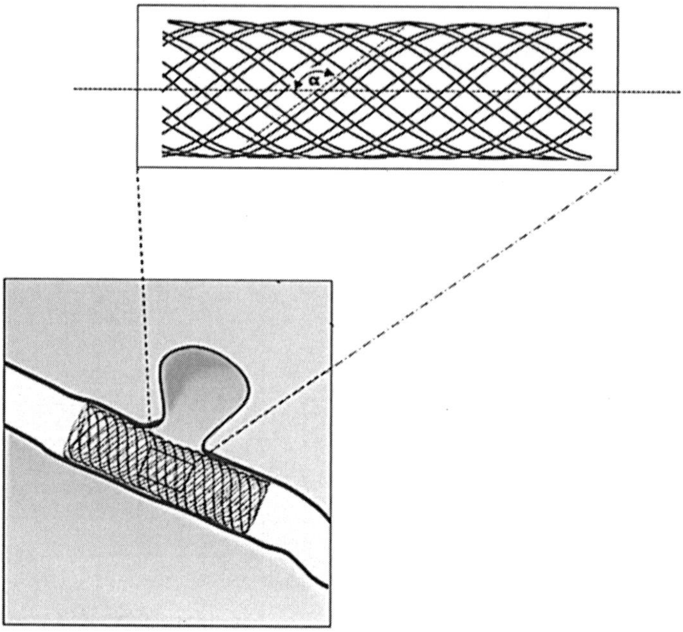

Figure 8. Structural geometry of a braided metallic stent (α = braid helix angle).

The research contribution towards improving braided stent mechanics is very limited but still worth mentioning. The concept of using a multilayer braid design was introduced to reduce porosity of the stent while maintaining the bending flexibility, which otherwise becomes too low in a high mesh density single layer stent to be able to conform well to the arterial wall. The use of multiple layer braided stent has been trialled in many clinical studies and is commercially available as CARDIATIS® multilayer stent [101,102,103]. Another new design concept was reported recently which is a single layer braided structure but with two different zones/sections along its length [104,105]. The central zone (70% of stent length) is a tightly braided high mesh density section while the margins have low mesh density. This allows reduced blood flow to the diseased portion of the blood vessel while maintaining flexibility in rest of the "non-functional" stent length. The commercial version of this design is available as E-volution® stent (Jotec GmbH, Germany). Marchand *et al.* tried to improve the radial compliance of a braided heart valve stent by optimising shape-setting parameters of the stent material (nitinol alloy) [106]. The modified heat treatment process was reported to improve radial expandability of the stent when tested in an *in vitro* test experiment.

3.4. Electrospinning

Since their introduction, Dacron® grafts (woven and knitted) have dominated the vascular graft market in large diameter (15–30 mm) blood vessel replacement. Dacron® grafts have also been used as replacements for small diameter (<6 mm) arteries (coronary, below-the-knee, tibial, and peroneal). However, there were a number of limitations identified such as owing to mismatched compliance, graft thrombosis, and anastomotic intimal hyperplasia (Table 4) [6,23,107,108,109,110]. One of the most commonly used strategies to address some of these limitations is by coating the synthetic grafts with protein such as collagen. As shown in Table 5, the general trend observed is that grafts coated with collagen, fibronectin or heparin demonstrated better cell attachment *in vitro* and improved patency *in vivo* [111,112,113,114].

The graft structure (porosity, fibre diameters, pore connectivity, surface area, and compliance) is important as it influences cell growth behaviour. This led to the development of a new textile fibre spinning technique known as electrospinning, which is capable of producing fibres to the scale of native collagen and elastin fibres. In an electrospinning process, a strong electric field is generated between a polymer solution (delivered through a syringe needle) and a metallic collector. When the voltage reaches a critical value, the charge overcomes the surface tension of the polymer solution drop at the needle tip and polymer jet is generated. While travelling towards the metallic collector the drawing force exerted by electric field and simultaneous evaporation of the solvent results in reduction in the diameter of the jet. The collected dry fibres form a nonwoven mesh of nanometre to micrometre diameter fibres (Figure 9). The process can be adjusted to control fibre diameter to some extent by varying the charge density and polymer solution concentration. The feature of producing very fine fibres increases the surface area significantly in electrospun meshes, which is a beneficial property for improving cell growth [116].

The structural construction of an electrospun fibre scaffold can closely mimic the natural extracellular matrix structure [117]. Electrospinning also enables the use of a wide range of natural and synthetic polymers [118], thus, increasing the possibility of matching both the biological and mechanical properties similar to an artery (Table 5) [119]. The electrospinning of natural proteins like collagen and elastin have demonstrated the usefulness of this technique to bring synthetic vascular grafts near to their biological counterparts [120,121]. From a structural design perspective also, electrospinning provides a range of possibilities in mimicking the multilayered construction of arterial wall [122,123,124,125]. The use of different polymeric components in individual layers provides more flexibility in controlling the graft mechanical response from each structural layer.

Table 4. Compliance and cumulative patency of different arterial grafts implanted in a small diameter (femoropopliteal artery) position. Adapted from [115]; PTFE, Polytetrafluoroethylene.

Graft Type	**Compliance (% mmHg × 10^{-2})**	**Structure**	**Patency % (30/180 days)**	**Patency % (1/2 years)**
Human femoral artery	5.9	–	–	–
Saphenous vein	4.4	Natural tissue	94/93	88/84
Umbilical vein	3.7	Natural tissue	97/93	83/80
PTFE	1.6	Extruded (non-textile)	85/81	60/42
Dacron®	1.9	Woven	88/72	65/42

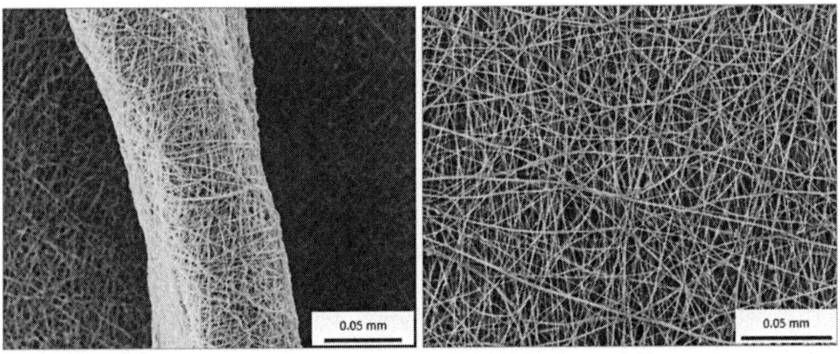

(a) (b)

Figure 9. The wall thickness (**a**) and surface (**b**) view of an electrospun mesh.

4. DISCUSSION

The woven graft development timeline suggests that this technology has shown biomechanical benefits of using new structural components but there are very limited attempts which focus on improvising conventional weave designs to suit the arterial site. Knitted structures appear to be a suitable candidate for vascular implant application owing to their inherent structural and design flexibility. Also, anisotropic elasticity property (axial > circumferential) in knitted structures is similar to that of the native artery, which is an advantage over woven structures. Currently, some of the latest developments in woven and knitted grafts use biological coatings to improved tissue growth and design modifications to match the implant site anatomy, while the lack of innovation from material and biomechanical aspect is evident in new commercial products (Table 6). The future products may require a combined input from advanced textile designing and biomechanics together in order to realise the full clinical potential of textile grafts. The helical arrangement of constituent filaments/wires give braided structures a design advantage over woven and knitted structures to mimic the helical geometry of collagen and elastin fibres in an artery. However, the focus on this aspect of braided stent improvement is mostly unrealised. Since, braided stents are deployed in their fully expanded state (helix angle approaching 90°), the low stress radial expansion property owing to helical geometry is completely lost (Figure 8). The use of elastic filaments and design of bio-component structures can be a future prospect for developing compliant braided stent devices. Expanded polytetrafluoroethylene (ePTFE) is an inert fluorocarbon polymer (stiffness: 0.5 GPa, tensile strength: 14–18 MPa), which is developed by heating, stretching, and extruding process resulting in a porous polymeric structure. Non-textile grafts made from ePTFE are currently used in the clinics as medium diameter grafts (7–9 mm) for peripheral vascular diseases. However, ePTFE grafts are not viable as conduits for small diameter (<6 mm) vessel replacement due to a high rate of occlusion. Although ePTFE is a chemically inert material, the structural stiffness and low radial compliance of these grafts contributes to their poor long-term patency (Table 4). There have been numerous investigations into endothelial cell seeding and surface functionalization of ePFTE grafts to improve their clinical performance as small diameter grafts. Current efforts have yet to produce a small diameter synthetic graft that is comparable to an autologous graft. Electrospinning technique has proven to be a promising option for small diameter grafts in many *in vitro* and animal studies. In comparison to established textile manufacturing methods (weaving, knitting, braiding), this technology is equipped with much higher levels of design flexibility in terms of material variety (natural, synthetic), structural heterogeneity (multi-layer, multi-component), and therapeutic ability (drug delivery). However, development of an off-the-shelf electrospun graft product capable of achieving rapid cell coverage with minimised risk of thrombosis, intimal hyperplasia, and mechanical failure has yet to be achieved.

Table 5. Biological responses in various types of synthetic vascular grafts created from different manufacturing techniques.

Technique	Material	Scaffold Dimensions	Biological Response	Mechanical Testing	Ref.
Knitted	PET	4 mm diameter	ECs: Better cell attachment was observed on precoated grafts in the following order: collagen (3.5×10^5 cells/cm^2) > fibrin (2.8×10^5 cells/cm^2) > fibronectin (2.4×10^5 cells/cm^2) > laminin = untreated (1.3×10^5 cells/cm^2)	–	[126]
Knitted	PET	4 mm diameter cut into fusiform patches of 5 cm length × 8 mm width	*In vivo*: Implanted in sheep for 4 weeks and intimal hyperplasia assessed. Fluoropolymer coated PET sealed with gelatin produced the least amount of hyperplasia followed by PTFE, carbonlined PTFE and gelatin sealed PET	–	[127]

Technique	Material	Scaffold Dimensions	Biological Response	Mechanical Testing	Ref.
Knitted	PET	–	*In vivo*: Randomised clinical trial of 209 patients (femoropopliteal bypass). Patency at 3 years of collagen-coated heparin bonded PET (54%) was better than PTFE (44%). No difference in patency at 5 years between PET and PTFE	–	[128]
Woven and Electrospinning	PET and PU	Flat	*In vitro* (HUVECs): All 3 materials showed better cell attachment when coated with Collagen Type I/III as compared to their uncoated counterparts. Cell coverage on coated PTFE (34.6%) > PET (19.6%) > PU (17.5%); When exposed to shear stress of 1 Pa for 1 h, cell retention was highest on coated PTFE. No difference was observed across the 3 materials when shear stress increased to 2 Pa	–	[129]
Electrospinning	PCL and PU	NS	*In vitro* (HUVECs): Good cell attachment (~ 61%). Cells produced proteins such as PECAM and vWF, indicating normal phenotype and function	–	[130]
Electrospinning	Silk fibroin	1.5 mm diameter	*In vivo*: Implanted in rats for 7 days to 6 months. No graft degradation or foreign body reaction observed. 95% of grafts remained unblocked. After 1 month implantation, ECs were observed on the luminal surface of the grafts. Cell coverage continue to increase at 6 months	–	[131]
Electrospinning	Silk	5 mm diameter	*In vitro*: ECs and SMCs attached and proliferated on the scaffold. ECs were observed on the surface of the scaffold while SMCs migrated into the scaffold	Tensile strength: 2.42 MPa; Elastic modulus: 2.45 MPa; Mean burst pressure: 811 mmHg	[132]
Electrospinning	PCL	2 mm diameter	*In vivo*: Implanted in rats for 24 weeks. No narrowing of the grafts (stenosis) in the PCL group. ECs coverage confluent at 12 weeks in the PCL group *vs.* ePTFE group (incomplete at 24 weeks).	–	[133]
Electrospinning	PCL	NS	*In vitro*: A confluent layer of oriented SMCs in the direction of aligned fibres after 7 days culture	–	[134]

Technique	Material	Scaffold Dimensions	Biological Response	Mechanical Testing	Ref.
Electrospinning	Collagen Type 1, elastin and PLGA	Tubular (4.75 mm diameter, 12 cm long)	*In vitro*: Mean of 72% for ECs and 82% for SMCs viability over 7-day culture period; *In vivo* (mice): No systemic or neurological toxicity, normal blood count, transient local inflammation at implantation site	Burst pressure: 1425 mm Hg; Compliance: 12%–14% for scaffold *vs.* 9% for native vessel; Maximal axial and circumferential strain: 40% strain	[119]
Electrospinning	Collagen and PCL	Tubular (4.75 mm diameter, 12 cm long)	ECs: Cytoskeletal organisation and focal adhesion via actin and vinculin staining respectively were better developed when cultured on smaller sized fibres; SMCs: Infiltration of cells into scaffolds with fibres > 1 μm in diameter during a 4-week culture	Tensile strength: Increasing fibre diameter (0.27 μm to 4.45 μm) decreased tensile strength from 3.15 MPa to 0.75 MPa; Elongation at break: Increased with increasing fibre diameter (90% for 0.27 μm fibres to 734% for 4.45 μm fibres); Maximum load: Decreased from 25.75 N (0.27 μm fibres) to 8.63 N (4.45 μm fibres)	[135]
Knitting	Elastin solubilised proteins and PET	Flat	HUVECs: No cytotoxicity, 43% cell attachment for elastin solubilised protein-PET *vs.* 94% for PET	–	[136]
Electrospinning	Elastin and PDO	6 mm diameter	*In vitro*: Fibroblasts cultured for 7 days on PDO:Elastin graft showed migration into the fibrous materials *vs.* no migration in PDO graft	PDO:Elastin ratio of 50:50 produced compliance that mimics native femoral artery	[137]
Electrospinning	PCL, PDO and Silk	NS	*In vitro*: Risk of clotting using human monocytes – PCL < Silk < PDO. The risk in PCL is comparable to ePTFE (currently used grafts in clinics)	–	[138]
Electrospinning	Chitosan and PVA	NS	*In vitro*: Good fibroblast growth was observed with continual proliferation up to 21 days	–	[139]

EC = endothelial cell; ePTFE = expanded polytetrafluoroethylene; HUVEC = human umbilical vein endothelial cell; NS = not specified; PCL = polycaprolactone; PDO = polydioxanone; PET = polyethylene terephthalate; PLGA = poly(lactic-co-glycolic acid); PTFE = polytetrafluoroethylene; PU = polyurethane; PVA = polyvinyl alcohol; SMC = smooth muscle cell.

Radial compliance is unarguably the most vital mechanical feature requirement from a synthetic vascular implant. In conjunction with compliance, non-linearity and anisotropy are two critical mechanical characteristics, which a vascular graft must also possess in order to successfully mimic natural artery mechanics. Non-linearity prevents the late development of pathological conditions related to graft failure at high pressures [2,3]. Anisotropic mechanical property prevents excessive stimulation of anastomotic region, while being compliant in radial direction to prevent any flow disturbances [125]. In an artery, both these properties are achieved owing to the multilayered heterogeneous structure of arterial wall, which a single-layer single-component isotropic textile structure is unable to mimic.

Table 6. A list of selected commercially available Dacron® grafts and their design features.

Structure	Material	Trade Name	Special Design Feature	Intended Improvement	Manufacturer	Application Area
Woven	Dacron® & PTFE	FUSION BIOLINE®	Two layer, Heparin coated	Improves patency and healing response	Macquet	Peripheral
	Dacron®	HEMASHIELD® PLATINUM	Multiple branched, Double velour, Collagen coated	Enhances healing response		Aortic
		VASCUTEK® GELWEAVE™ Pre-curved	Pre-curved design	Matches aortic arch anatomy	Terumo	
		VASCUTEK® SIENA™	Extended trunk and collar design	Suits hybrid surgery procedures		
		VASCUTEK® GELWEAVE™ Plexus	Multiple branched, Gelatin impregnated	Suits complete aortic arch surgery		
		VASCUTEK® GELWEAVE™ Valsalva	Extended skirt design	Matches aortic root anatomy		
		VASCUTEK® GELWEAVE™ Ante-Flo	Extra branch	Reduces surgery time, Lowers risk of neurological deficits		
		E-VITA™ OPEN PLUS	Extended stented trunk	Suits hybrid surgery procedures	Jotec	
Knitted		INTEGRAD® SILVER	Silver impregnated	Reduces graft infection	Macquet	Aortic, Peripheral
		HEMAGARD® Ultrathin	Collagen coated, Wall thickness = 0.35 mm	Improves healing response, Reduces dilatation, Increases suture strength		Aortic
		HEMASHIELD® GOLD	Collagen coated, Double velour surface	Improves healing response, Reduces dilatation, Increases suture strength		
		VASCUTEK® GELSEAL™	Gelatin impregnated	Improves healing	Terumo	Aortic, Peripheral
		VASCUTEK® GELSOFT™ ERS	Gelatin impregnated, Radially supported	Improves healing and handling		
		FLOWNIT BIOSEAL®	Texturised yarn, Collagen impregnated	Low dilatation, Enhances tissue incorporation	Jotec	Aortic

5. CONCLUSIONS

This paper reviewed the design aspect of medical textiles (woven, knitted, braided, electrospun) intended for vascular implant applications. The three dimensional structure of an arterial wall and its unique mechanical properties (anisotropy, non-linearity, compliance, viscoelasticity) have been widely researched and reported. However, these features remain widely unconsidered while designing synthetic vascular implants. Therefore, the vascular implants used currently in clinics serve the function of a rigid non-distensible conduit but lack the ability to revive lost biomechanical function of the diseased artery. Currently available textile vascular implants are not significantly different from those introduced six decades ago. Their structural geometry is analogous to traditional textile structures rather than to an arterial vessel. This difference is highlighted when late clinical complications arise from behavioural mismatch at the artery-implant anastomosis. These observations ultimately raise the importance of understanding the structure and biomechanics of an artery before adapting a textile structure from its conventional application area to a biological environment consisting of complex structure and functions such as the blood vessel. A synthetic vascular implant with structural characteristics that closely resemble a native artery will present with less complications, which, in turn, translates to a longer-lasting implant.

AUTHOR CONTRIBUTIONS

C.S., C.S.W. and X.W. planned the manuscript. C.S. and C.S.W. wrote the manuscript. X.W. commented on and modified the manuscript.

CONFLICTS OF INTEREST

The authors declare no conflict of interest.

REFERENCES

1. Roach, M.R.; Burton, A.C. The reason for the shape of the distensibility curves of arteries. *Can. J. Biochem. Physiol.* **1957**, *35*, 681–690.

2. Shadwick, R.E. Elasticity in Arteries. *Am. Sci.* **1998**, *86*, 535–541.

3. Shadwick, R.E. Mechanical design in arteries. *J. Exp. Biol.* **1999**, *202*, 3305–3313.

4. Abbott, W.M.; Megerman, J.; Hasson, J.E.; L'Italien, G.; Warnock, D.F. Effect of compliance mismatch on vascular graft patency. *J. Vasc. Surg.* **1987**, *5*, 376–382.

5. Weston, M.W.; Rhee, K.; Tarbell, J.M. Compliance and diameter mismatch affect the wall shear rate distribution near an end-to-end anastomosis. *J. Biomech.* **1996**, *29*, 187–198.

6. Chlupác, J.; Filová, E.; Bačáková, L. Blood vessel replacement: 50 years of development and tissue engineering paradigms in vascular surgery. *Physiol. Res.* **2009**, *58*, 119–140.

7. Kannan, R.Y.; Salacinski, H.J.; Butler, P.E.; Hamilton, G.; Seifalian, A.M. Current status of prosthetic bypass grafts: A review. *J. Biomed. Mater. Res. B.* **2005**, *74*, 570–581.

8. Salacinski, H.J.; Goldner, S.; Giudiceandrea, A.; Hamilton, G.; Seifalian, A.M.; Edwards, A.; Carson, R.J. The mechanical behavior of vascular grafts: A review. *J. Biomater. Appl.* **2001**, *15*, 241–278.

9. Pourdeyhimi, B.; Text, C. A review of structural and material properties of vascular grafts. *J. Biomater. Appl.* **1987**, *2*, 163–204.

10. Pourdeyhimi, B.; Wagner, D. On the correlation between the failure of vascular grafts and their structural and material properties: A critical analysis. *J. Biomed. Mater. Res.* **1986**, *20*, 375–409.

11. Silver, F.; Snowhill, P.; Foran, D. Mechanical behavior of vessel wall: A comparative study of aorta, vena cava, and carotid artery. *Ann. Biomed. Eng.* **2003**, *31*, 793–803.

12. Fischer, G.M.; Llaurado, J.G. Collagen and elastin content in canine arteries selected from functionally different vascular beds. *Circ. Res.* **1966**, *19*, 394–399.

13. Kassab, G.S. Biomechanics of the cardiovascular system: The aorta as an illustratory example. *J. R. Soc. Interface* **2006**,*3*, 719–740.

14. Belz, G.G. Elastic properties and Windkessel function of the human aorta. *Cardiovasc. Drugs Ther.* **1995**, *9*, 73–83.

15. Greenfield, J.C.; Patel, D.J. Relation between pressure and diameter in the ascending aorta of man. *Circ. Res.* **1962**, *10*, 778–781.

16. Lillie, M.A.; Shadwick, R.E.; Gosline, J.M. Mechanical anisotropy of inflated elastic tissue from the pig aorta. *J. Biomech.* **2010**, *43*, 2070–2078.

17. Zhang, W.; Liu, Y.; Kassab, G.S. Viscoelasticity reduces the dynamic stresses and strains in the vessel wall: Implications for vessel fatigue. *Am. J. Physiol.* **2007**, *293*, H2355–H2360.

18. Blakemore, A.H.; Voorhees, A.B. The use of tubes constructed from vinyon "N" cloth in bridging arterial defects—Experimental and clinical. *Ann. Surg.* **1954**, *140*, 324–333.

19. Voorhees, A.B.J.; Jaretzki, A.I.; Blakemore, A.H. The use of tubes constructed from Vinyon "N" cloth in bridging arterial defects: A preliminary report. *Ann. Surg.* **1952**, *135*, 332–336.

20. Harris, E.J.; Shumacker, H.B., Jr.; Siderys, H.; Moore, T.C.; Grice, P.F. Pliable plastic aortic grafts: Experimental comparison of a number of materials. *AMA Arch. Surg.* **1955**, *71*, 449–459.

21. Hufnagel, C.A.; Rabil, P. Replacement of arterial segments, utilizing flexible orlon prostheses. *AMA Arch. Surg.* **1955**,*70*, 105–110.

22. Deterling, R.A., Jr.; Bhonslay, S.B. Use of vessel grafts and plastic prostheses for relief of superior vena caval obstruction. *Surgery.* **1955**, *38*, 1008–1026.

23. Harrison, J.H. Synthetic materials as vascular prostheses. I. A comparative study in small vessels of nylon, dacron, orlon, ivalon sponge and teflon. *Am. J. Surg.* **1958**, *95*, 3–15.

24. Julian, O.C.; Deterling, R.A., Jr.; Su, H.H.; Dye, W.S.; Belio, M.L. Dacron tube and bifurcation arterial prostheses produced to specification. Experimental and clinical use. *Surgery* **1957**, *41*, 50–61.

25. Knox, G.; Begg, C.F. Evaluation of the need for porosity in synthetic arterial prostheses. A case report and evaluation of dacron aortic grafts in dogs. *Surgery* **1957**, *42*, 922–927.

26. Harrison, J.H. Synthetic materials as vascular prostheses: II. A comparative study of nylon, dacron, orlon, ivalon sponge and teflon in large blood vessels with tensile strength studies. *Am. J. Surg.* **1958**, *95*, 16–24.

27. Tremblay, D.; Zigras, T.; Cartier, R.; Leduc, L.; Butany, J.; Mongrain, R.; Leask, R.L. A comparison of mechanical properties of materials used in aortic arch reconstruction. *Ann. Thorac. Surg.* **2009**, *88*, 1484–1491.

28. Lantelme, P.; Dzudie, A.; Milon, H.; Bricca, G.; Legedz, L.; Chevalier, J.M.; Feugier, P. Effect of abdominal aortic grafts on aortic stiffness and central hemodynamics. *J. Hypertens.* **2009**, *27*, 1268–1276.

29. Kim, S.Y.; Hinkamp, T.J.; Jacobs, W.R.; Lichtenberg, R.C.; Posniak, H.; Pifarré, R. Effect of an inelastic aortic synthetic vascular graft on exercise hemodynamics. *Ann. Thorac. Surg.* **1995**, *59*, 981–989.

30. Obrien, T.; Morris, L.; McGloughlin, T. Evidence suggests rigid aortic grafts increase systolic blood pressure: Results of a preliminary study. *Med. Eng. Phys.* **2008**, *30*, 109–115.

31. Fung, Y.C. *Biomechanics: Mechanical Properties of Living Tissues*; Springer: New York, NY, USA, 2013.

32. Tai, N.R.; Salacinski, H.J.; Edwards, A.; Hamilton, G.; Seifalian, A.M. Compliance properties of conduits used in vascular reconstruction. *Br. J. Surg.* **2000**, *87*, 1516–1524.

33. Kidson, I.G. The effect of wall mechanical properties on patency of arterial grafts. *Ann. R. Coll. Surg. Engl.* **1983**, *65*, 24–29.

34. Anderson, K.; Seyam, A. Developing seamless shaped woven medical products. *J. Med. Eng. Technol.* **2004**, *28*, 110–116.

35. Kotsuka, Y.; Ezure, M.; Tanaka, K.; Takamoto, S.; Kuwahara, K.; Ikeda, H. Ultrathin-wall vascular grafts for endovascular surgery. *J. Artif. Organs* **2002**, *5*, 108–112.

36. De Paulis, R.; Scaffa, R.; Maselli, D.; Salica, A.; Bellisario, A.; Weltert, L. A third generation of ascending aorta dacron graft: Preliminary experience. *Ann. Thorac. Surg.* **2008**, *85*, 305–309.

37. Abdessalem, S.B.; Durand, B.; Akesbi, S.; Chakfe, N. Blood flow in a polyester textile vascular prosthesis: Experimental and numerical study. *Text. Res. J.* **2001**, *71*, 178–183.

38. Hajjaji, R.; Abdessalem, S.B.; Ganghoffer, J.F. The influence of textile vascular prosthesis crimping on graft longitudinal elasticity and flexibility. *J. Mech. Behav. Biomed. Mat.* **2012**, *16*, 73–80.

39. Moghe, A.K.; Gupta, B.S. Small-diameter blood vessels by weaving: Prototyping and modelling. *J. Text. Inst.* **2008**, *99*, 467–477.

40. Hasegawa, M.; Azuma, T. Mechanical properties of synthetic arterial grafts. *J. Biomech.* **1979**, *12*, 509–517.

41. Wagner, M.; Reul, G.; Teresi, J.; Kayser, K.L. Experimental observations on a new and inherently elastic material for sutures and vascular prostheses: Lycra. *Am. J. Sur.* **1966**, *111*, 838–841.

42. Szilagyi, D.; France, L.C.; Smith, R.F.; Whitcomb, J.G. The clinical use of an elastic dacron prosthesis. *AMA Arch. Surg.***1958**, *77*, 538–551.

43. Szilagyi, D.; Whitcomb, J.G.; Shonnard, C.P. Replacement of long and narrow arterial segments: II. Experimental studies with an elastic ("Helanca") seamless woven nylon prosthesis. *AMA Arch. Surg.* **1957**, *74*, 944–953.

44. Gupta, B.S.; Kasyanov, V.A. Biomechanics of human common carotid artery and design of novel hybrid textile compliant vascular grafts. *J. Biomed. Mater. Res.* **1997**, *34*, 341–349.

45. Rosenberg, N.; Simpson, A.N.; Brown, R.E. A circumferentially elastic arterial prosthesis: Three-year studies of a dacron-spandex graft in the dog. *J. Surg. Res.* **1983**, *34*, 7–16.

46. Chen, Y.; Ding, X.; Li, Y.; King, M.W.; Gao, J.; Zhao, X. A bilayer prototype woven vascular prosthesis with improved radial compliance. *J. Text. Inst.* **2011**, *103*, 106–111.

47. Sonoda, H.; Takamizawa, K.; Nakayama, Y.; Yasui, H.; Matsuda, T. Small-diameter compliant arterial graft prosthesis: Design concept of coaxial double tubular graft and its fabrication. *J. Biomed. Mater. Res.* **2001**, *55*, 266–276.

48. Sanger, P.W.; Taylor, F.H.; McCall, R.E.; Duchesne, R.; Lepage, G. Seamless synthetic arterial grafts; preliminary report on experimental and clinical experiences. *J. Am. Med. Assoc.* **1956**, *160*, 1403–1404.

49. De Bakey, M.; Cooley, D.A. Successful resection of aneurysm of distal aortic arch and replacement by graft. *J. Am. Med. Assoc.* **1954**, *155*, 1398–1403.

50. De Bakey, M.E.; Cooley, D.A.; Crawford, E.; Morris, G.C., Jr. Clinical application of a new flexible knitted dacron arterial substitute. *AMA Arch. Surg.* **1958**, *77*, 713–724.

51. Chakfe, N.; Dieval, F.; Wang, L.; Thaveau, F.; Rinckenbach, S.; Edah-Tally, S.; Mathieu, D.; Le Magnen, J.F.; Riepe, G.; Kretz, J.G.; Durand, B. *In vitro* approach to the dilative behavior of knitted vascular prosthetic grafts. *Ann. Vasc. Surg.* **2008**, *22*, 402–411.

52. Chakfé, N.; Riepe, G.; Dieval, F.; Le Magnen, J.F.; Wang, L.; Urban, E.; Beaufigeau, M.; Durand, B.; Imig, H.; Kretz, J.G. Longitudinal ruptures of polyester knitted vascular prostheses. *J. Vasc. Surg.* **2001**, *33*, 1015–1021.

53. Dieval, F.; Chakfé, N.; Wang, L.; Riepe, G.; Thaveau, F.; Heintz, C.; Mathieu, D.; Le Magnen, J.F.; Kretz, J.G.; Durand, B. Mechanisms of rupture of knitted polyester vascular prostheses: An *in vitro* analysis of virgin prostheses. *Eur. J. Vasc. Endovasc. Surg.* **2003**, *26*, 429–436.

54. Ichikawa, H.; Ishikawa, S.; Ohtaki, A.; Otani, Y.; Sakata, K.; Takahashi, T.; Aizaki, M.; Sato, Y.; Morishita, Y. Clinical experience with a knitted Dacron graft impregnated with gelatin (Gelseal Triaxial®). *Artif. Organs* **1996**, *5*, 59–63.

55. Kudo, F.A.; Nishibe, T.; Miyazaki, K.; Flores, J.; Yasuda, K. Albumin-coated knitted Dacron aortic prostheses. Study of prosperative inflammatory reactions. *Int. Angiol.* **2002**, *21*, 214–217.

56. Marais, Y.; Guidoin, R.; Deng, X.; King, M.W.; Martin, L.; Roy, R. The Dialine II graft: A new collagen-impregnated warp-knitted polyester arterial prosthesis. *Ann. Vasc. Surg.* **1997**, *11*, 133–140.

57. Mary, C.; Marois, Y.; King, M.W.; Hong, T.; Laroche, G.; Douville, Y.; Martin, L.; Guidoin, R. *In vitro* and *in vivo* studies of a polyester arterial prosthesis with a warp-knitted sharkskin structure. *J. Biomed. Mater. Res.* **1997**, *35*, 459–472.

58. Meister, R.H.; Schweiger, H.; Lang, W. Knitted double-velour Dacron prostheses in aortobifemoral position—Long-term performance of different coating materials. *Vasa J. Vasc. Disease.* **1998**, *27*, 236–239.

59. Phaneuf, M.D.; Dempsey, D.J.; Bide, M.J.; Quist, W.C.; LoGerfo, F.W. Coating of Dacron vascular grafts with an ionic polyurethane: a novel sealant with protein binding properties. *Biomaterials* **2001**, *22*, 463–469.

60. Sirry, M.S.; Zilla, P.; Franz, T. A computational study of structural designs for a small-diameter composite vascular graft promoting tissue regeneration. *Cardiovasc. Eng. Technol.* **2010**, *1*, 269–281.

61. Wang, F.J.; Mohammed, A.; Li, C.J.; Wang, L. Promising poly(ε-caprolactone) composite reinforced with weft-knitted polyester for small-diameter vascular graft application. *Adv. Mater. Sci. Eng.* **2014**, *2014*, 1–9.

62. Xu, W.; Zhou, F.; Ouyang, C.; Ye, W.; Yao, M.; Xu, B. Mechanical properties of small-diameter polyurethane vascular grafts reinforced by weft-knitted tubular fabric. *J. Biomed. Mater. Res.* **2010**, *92*, 1–8.

63. Yagi, T.; Sato, M.; Nakazawa, Y.; Tanaka, K.; Sata, M.; Itoh, K.; Takagi, Y.; Asakura, T. Preparation of double-raschel knitted silk vascular grafts and evaluation of short-term function in a rat abdominal aorta. *J. Artif. Organs* **2011**, *14*, 89–99.

64. Yang, H.; Zhu, G.; Zhang, Z.; Wang, Z.; Fang, J.; Xu, W. Influence of weft-knitted tubular fabric on radial mechanical property of coaxial three-layer small-diameter vascular graft. *J. Biomed. Mater. Res. B* **2012**, *100*, 342–349.

65. Freitas, A.F.D.P.; de Araujo, M.D.; Zu, W.W.; Fangueiro, R.M.E. Development of weft-knitted and braided polypropylene stents for arterial implant. *J. Text. Inst.* **2010**, *101*, 1027–1034.

66. Li, G.; Li, Y.; Lan, P.; He, X.; Hu, H. Polydioxanone weft-knitted intestinal stents: fabrication and mechanics optimization. *Text. Res. J.* **2013**, *83*, 2129–2141.

67. Moodley, L.; Franz, T.; Human, P.; Wolf, M.F.; Bezuidenhout, D.; Scherman, J.; Zilla, P. Protective constriction of coronary vein grafts with knitted nitinol. *Eur. J. Cardiothorac. Surg.* **2013**, *44*, 64–71.

68. Weinandy, S.; Rongen, L.; Schreiber, F.; Cornelissen, C.; Flanagan, T.C.; Mahnken, A.; Gries, T.; Schmitz-Rode, T.; Jockenhoevel, S. The BioStent: Novel concept for a viable stent structure. *Tissue Eng. A* **2012**, *18*, 1818–1826.

69. Zilla, P.; Moodley, L.; Wolf, M.F.; Bezuidenhout, D.; Sirry, M.S.; Rafiee, N.; Lichtenberg, W.; Black, M.; Franz, T. Knitted nitinol represents a new generation of constrictive external vein graft meshes. *J. Vasc. Surg.* **2011**, *54*, 1439–1450.

70. Singh, C.; Wang, X. A biomimetic approach for designing stent-graft structures: Caterpillar cuticle as design model. *J. Mech. Behav. Biomed. Mater.* **2014**, *30*, 16–29.

71. Singh, C.; Wang, X. A new design concept for knitted external vein-graft support mesh. *J. Mech. Behav. Biomed. Mater.* **2015**, *48*, 125–133.

72. Self, M.M.; Cooley, D.A.; Debakey, M.E.; Creech, O., Jr. The use of braided nylon tubes for aortic replacement. *Ann. Surg.* **1955**, *142*, 836–843.

73. Edwards, W.S.; Tapp, J.S. Braided textiles tubes as arterial grafts. *Surgery* **1995**, *38*, 61–67.

74. Dale, W.; Niguidula, F.N. Study of elasticized dacron as arterial prosthesis: Experimental comparison with other plastics, homologous arteries, and autogenous veins. *AMA Arch. Surg.* **1959**, *78*, 246–259.

75. Whittlesey, R.H.; Sekerak, N.M.; Blunt, J.W., Jr.; Hudack, S.S. Observations of elastic synthetic fabric sleeves as vascular prostheses. *AMA Arch. Surg.* **1956**, *73*, 440–454.

76. Parodi, J.C.; Palmaz, J.C.; Barone, H.D. Transfemoral intraluminal graft implantation for abdominal aortic aneurysms. *Ann. Vasc. Surg.* **1991**, *5*, 491–499.

77. Wilson, G.; Klement, P.; Kato, Y.P.; Martin, J.B.; Khan, I.J.; Alcime, R.; Dereume, J.-P.; Macgregor, D.C.; Pinchuk, L. A self-expanding bifurcated endovascular graft for abdominal Aortic aneurysm repair an initial study in a canine model. *ASAIO J.* **1996**, *42*, M386–M393.

78. Georg, Y.; Settembre, N.; Marchand, C.; Lejay, A.; Thaveau, F.; Durand, B.; Chakfe, N. Poor long-term stability of the Corvita abdominal stentgraft. *Eur. J. Vasc. Endovasc. Surg.* **2014**, *47*, 160–163.

79. Raillat, C.; Rousseau, H.; Joffre, F.; Roux, D. Treatment of iliac artery stenoses with the Wallstent endoprosthesis. *Am. J. Roentgenol.* **1990**, *154*, 613–616.

80. Kallmes, D.F.; Ding, Y.H.; Dai, D.; Kadirvel, R.; Lewis, D.A.; Cloft, H.J. A new endoluminal, flow-disrupting device for treatment of saccular aneurysms. *Stroke* **2007**, *38*, 2346–2352.

81. Kallmes, D.F.; Ding, Y.H.; Dai, D.; Kadirvel, R.; Lewis, D.A.; Cloft, H.J. A Second-Generation, Endoluminal, Flow-Disrupting Device for Treatment of Saccular Aneurysms. *Am. J. Neuroradiol.* **2009**, *30*, 1153–1158.

82. Lieber, B.; Stancampiano, A.; Wakhloo, A. Alteration of hemodynamics in aneurysm models by stenting: Influence of stent porosity. *Ann. Biomed. Eng.* **1997**, *25*, 460–469.

83. Mallik, A.S.; Nuss, K.; Kronen, P.W.; Klein, K.; Karol, A.; von Rechenberg, B.; Rüfenacht, D.A.; Wanke, I.; Kulcsár, Z. A new-generation, low-permeability flow diverting device for treatment of saccular aneurysms. *Eur. Radiol.* **2014**, *24*, 12–18.

84. Alderson, H.; Zamir, M. Effects of stent stiffness on local haemodynamics with particular reference to wave reflections. *J. Biomech.* **2004**, *37*, 339–348.

85. Nii, K.; Tsutsumi, M.; Aikawa, H.; Hamaguchi, S.; Etou, H.; Sakamoto, K.; Kazekawa, K. Incidence of hemodynamic depression after carotid artery stenting using different self-expandable stent types. *Neurol. Med. Chir.* **2011**, *51*, 556–560.

86. Tanaka, N.; Martin, J.B.; Tokunaga, K.; Abe, T.; Uchiyama, Y.; Hayabuchi, N.; Berkefeld, J.; Rüfenacht, D.A. Conformity of carotid stents with vascular anatomy: Evaluation in carotid Models. *Am. J. Neuroradiol.* **2004**, *25*, 604–607.

87. Hyun Kim, J.; Jin Kang, T.; Yu, W.-R. Simulation of mechanical behavior of temperature-responsive braided stents made of shape memory polyurethanes. *J. Biomech.* **2010**, *43*, 632–643.

88. Irsale, S.; Adanur, S. Design and characterization of polymeric stents. *J. Ind. Text.* **2006**, *35*, 189–200.

89. Lin, J.H.; Wen, S.P.; Lou, C.W.; Ueng, K.C. Compression behaviors of stainless steel braided coronary stents and polyvinyl alcohol/stainless steel braided coronary stents. *Adv. Mater. Res.* **2014**, *910*, 274–278.

90. Mikkonen, J.; Uurto, I.; Isotalo, T.; Kotsar, A.; Tammela, T.L.J.; Talja, M.; Salenius, J.P.; Törmälä, P.; Kellomäki, M. Drug-eluting bioabsorbable stents—An *in vitro* study. *Acta Biomater.* **2009**, *5*, 2894–2900.

91. Nuutinen, J.P.; Välimaa, T.; Clerc, C.; Törmälä, P. Mechanical properties and *in vitro* degradation of bioresorbable knitted stents. *J. Biomater. Sci. Polym. Ed.* **2002**, *13*, 1313–1323.

92. Rebelo, R.; Vila, N.; Fangueiro, R.; Carvalho, S.; Henriques, M. Development of braided fiber-based stents. *Stud. Health Technol. Inform.* **2014**, *207*, 135–144.

93. Ueng, K.-C.; Wen, S.-P.; Lou, C.-W.; Lin, J.-H. Braiding structure stability and section treatment evaluations of braided coronary stents made of stainless steel and bio-absorbable polyvinyl alcohol via a braiding technique. *Fib. Polym.* **2015**, *16*, 675–684.

94. van der Giessen, W.J.; Slager, C.J.; Van Beusekom, H.M.M.; Van Ingen Schenau, D.S.; Huijts, R.A.; Schuurbiers, J.C.; de Klein, W.J.; Serruys, P.W.; Verdouw, P.D. Development of a Polymer endovascular prosthesis and its implantation in porcine arteries. *J. Interv. Cardiol.* **1992**, *5*, 175–186.

95. Wang, X.; Li, Q.; Hu, X.; Ma, L.; You, C.; Zheng, Y.; Sun, H.; Han, C.; Gao, C. Fabrication and characterization of poly(L-lactide-co-glycolide) knitted mesh-reinforced collagen-chitosan hybrid scaffolds for dermal tissue engineering. *J. Mech. Behav. Biomed. Mater.* **2012**, *8*, 204–215.

96. Yuksekkaya, M.E.; Adanur, S. Analysis of polymeric braided tubular structures intended for medical applications. *Text. Res. J.* **2009**, *79*, 99–109.

97. De Beule, M.; Van Cauter, S.; Mortier, P.; Van Loo, D.; Van Impe, R.; Verdonck, P.; Verhegghe, B. Virtual optimization of self-expandable braided wire stents. *Med. Eng. Phys.* **2009**, *31*, 448–453.

98. Kim, J.H.; Kang, T.J.; Yu, W.R. Mechanical modeling of self-expandable stent fabricated using braiding technology. *J. Biomech.* **2008**, *41*, 3202–3212.

99. Makoyeva, A.; Bing, F.; Darsaut, T.E.; Salazkin, I.; Raymond, J. The varying porosity of braided self-expanding stents and flow diverters: An experimental study. *Am. J. Neuroradiol.* **2013**, *34*, 596–602.

100. Sui, W.; Jiang, J.; Chen, N. Effects of braiding parameters on radial compressive property of intraluminal stents. *Appl. Mech. Mater.* **2013**, *268*, 213–216.

101. Balderi, A.; Antonietti, A.; Pedrazzini, F.; Ferro, L.; Leotta, L.; Peano, E.; Grosso, M. Treatment of a hepatic artery aneurysm by endovascular exclusion using the multilayer cardiatis stent. *Cardiovasc. Intervent. Radiol.* **2010**, *33*, 1282–1286.

102. Henry, M.; Polydorou, A.; Frid, N.; Gruffaz, P.; Cavet, A.; Henry, I.; Hugel, M.; Rüfenacht, D.A.; Augsburger, L.; De Beule, M.; *et al.* Treatment of renal artery aneurysm with the multilayer stent. *J. Endovasc. Ther.* **2008**, *15*, 231–236.

103. Müller-Eschner, M.; Kortes, N.; Rehnitz, C.; Sumkauskaite, M.; Rengier, F.; Böckler, D.; Kauczor, H.-U.; Radeleff, B. Endovascular repair of a complex splenic artery aneurysm using a multilayer stent. *Cardiovasc. Intervent. Radiol.* **2015**,*38*, 494–497.

104. Ahlhelm, F.; Kaufmann, R.; Ahlhelm, D.; Ong, M.; Roth, C.; Reith, W. Carotid artery stenting using a novel self-expanding braided nickel–titanium stent: Feasibility and safety porcine trial. *Cardiovasc. Intervent. Radiol.* **2009**, *32*, 1019–1027.

105. Ahlhelm, F.; Roth, C.; Kaufmann, R.; Schulte-Altedorneburg, G.; Romeike, B.M.; Reith, W. Treatment of wide-necked intracranial aneurysms with a novel self-expanding two-zonal endovascular stent device. *Neuroradiology* **2007**, *49*, 1023–1028.

106. Marchand, C.; Heim, F.; Durand, B.; Chafke, N. Nitinol stent for percutaneous heart valve implantation: Material shape setting. *Mater. Manuf. Process.* **2011**, *26*, 181–187.

107. Massell, T.B.; Heringman, E.; Greenstone, S.M. Woven Dacron and woven Teflon prostheses: Use for small artery replacement. *Arch. Surg.* **1962**, *84*, 73–79.

108. Cate, W.R., Jr. Bypass grafts in small arteries: An experimental comparison of freeze-dried arterial homografts, autogenous vein and crimped dacron. *Am. Surg.* **1959**, *25*, 523–527.

109. Desai, M.; Seifalian, A.M.; Hamilton, G. Role of prosthetic conduits in coronary artery bypass grafting. *Eur. J. Cardiothorac. Surg.* **2011**, *40*, 394–398.

110. Seifalian, A.M.; Tiwari, A.; Hamilton, G.; Salacinski, H.J. Improving the clinical patency of prosthetic vascular and coronary bypass grafts: The role of seeding and tissue engineering. *Artif. Organs.* **2002**, *26*, 307–320.

111. Burkel, W.E.; Vinter, D.W.; Ford, J.W.; Kahn, R.H.; Graham, L.M.; Stanley, J.C. Sequential studies of healing in endothelial seeded vascular prostheses: Histologic and ultrastructure characteristics of graft incorporation. *J. Surg. Res.* **1981**, *30*, 305–324.

112. Fujita, Y.; Wu, M.H.D.; Ishida, A.; Shi, Q.; Walker, M.; Hammond, W.P.; Sauvage, L.R. Accelerated healing of Dacron grafts seeded by preclotting with autologous bone marrow blood. *Ann. Vasc. Surg.* **1999**, *13*, 402–412.

113. Pasquinelli, G.; Freyrie, A.; Preda, P.; Curti, T.; D'Addato, M.; Laschi, R. Healing of prosthetic arterial grafts. *Scanning Microsc.* **1990**, *4*, 351–362.

114. Tolan, M.; Wells, F.; Kendall, S.; Large, S.; Wallwork, J. Clinical experience with a collagen impregnated woven Dacron graft. *J. Cardiovasc. Surg.* **1995**, *36*, 323–327.

115. Walden, R.; L'Italien, G.J.; Megerman, J.; Abbott, W.M. Matched elastic properties and successful arterial grafting.*Arch. Surg.* **1980**, *115*, 1166–1169.

116. Li, W.-J.; Laurencin, C.T.; Caterson, E.J.; Tuan, R.S.; Ko, F.K. Electrospun nanofibrous structure: A novel scaffold for tissue engineering. *J. Biomed. Mater. Res.* **2002**, *60*, 613–621.

117. Xu, C.Y.; Inai, R.; Kotaki, M.; Ramakrishna, S. Aligned biodegradable nanofibrous structure: A potential scaffold for blood vessel engineering. *Biomaterials.* **2004**, *25*, 877–886.

118. Agarwal, S.; Wendorff, J.H.; Greiner, A. Use of electrospinning technique for biomedical applications. *Polymer* **2008**,*49*, 5603–5621.

119. Stitzel, J.; Liu, J.; Lee, S.J.; Komura, M.; Berry, J.; Soker, S.; Lim, G.; Van Dyke, M.; Czerw, R.; Yoo, J.J.; Atala, A. Controlled fabrication of a biological vascular substitute. *Biomaterials* **2006**, *27*, 1088–1094.

120. Buttafoco, L.; Kolkman, N.G.; Engbers-Buijtenhuijs, P.; Poot, A.A.; Dijkstra, P.J.; Vermes, I.; Feijen, J. Electrospinning of collagen and elastin for tissue engineering applications. *Biomaterials* **2006**, *27*, 724–734.

121. Wong, C.; Liu, X.; Xu, Z.; Lin, T.; Wang, X. Elastin and collagen enhances electrospun aligned polyurethane as scaffolds for vascular graft. *J. Mater. Sci. Mater. Med.* **2013**, *24*, 1865–1874.]

122. McClure, M.J.; Simpson, D.G.; Bowlin, G.L. Tri-layered vascular grafts composed of polycaprolactone, elastin, collagen, and silk: Optimization of graft properties. *J. Mech. Behav. Biomed. Mater.* **2012**, *10*, 48–61.

123. McClure, M.J.; Sell, S.A.; Simpson, D.G.; Walpoth, B.H.; Bowlin, G.L. A three-layered electrospun matrix to mimic native arterial architecture using polycaprolactone, elastin, and collagen: A preliminary study. *Acta Biomater.* **2010**, *6*, 2422–2433.

124. Soletti, L.; Hong, Y.; Guan, J.; Stankus, J.J.; El-Kurdi, M.S.; Wagner, W.R.; Vorp, D.A. A bilayered elastomeric scaffold for tissue engineering of small diameter vascular grafts. *Acta Biomater.* **2010**, *6*, 110–122.

125. Bonani, W.; Maniglio, D.; Motta, A.; Tan, W.; Migliaresi, C. Biohybrid nanofiber constructs with anisotropic biomechanical properties. *J. Biomed. Mater. Res. B* **2011**, *96*, 276–286.

126. Anderson, J.S.; Price, T.M.; Hanson, S.R.; Harker, L.A. *In vitro* endothelialization of small-caliber vascular grafts.*Surgery* **1987**, *101*, 577–586.

127. Ao, P.Y.; Hawthorne, W.J.; Vicaretti, M.; Fletcher, J.P. Development of intimal hyperplasia in six different vascular prostheses. *Eur. J. Vasc. Endovasc. Surg.* **2000**, *20*, 241–249.

128. Devine, C.; McCollum, C. Heparin-bonded Dacron or polytetrafluorethylene for femoropopliteal bypass: five-year results of a prospective randomized multicenter clinical trial. *J. Vasc. Surg.* **2004**, *40*, 924–931.

129. Feugier, P.; Black, R.A.; Hunt, J.A.; How, T.V. Attachment, morphology and adherence of human endothelial cells to vascular prosthesis materials under the action of shear stress. *Biomaterials* **2005**, *26*, 1457–1466.

130. Williamson, M.R.; Black, R.; Kielty, C. PCL-PU composite vascular scaffold production for vascular tissue engineering: attachment, proliferation and bioactivity of human vascular endothelial cells. *Biomaterials* **2006**, *27*, 3608–3616.

131. Bergmeister, H.; Grasl, C.; Walter, I.; Plasenzotti, R.; Stoiber, M.; Schreiber, C.; Losert, U.; Weigel, G.; Schima, H. Electrospun small-diameter polyurethane vascular grafts: ingrowth and differentiation of vascular-specific host cells.*Artif. Organs* **2012**, *36*, 54–61.

132. Soffer, L.; Wang, X.; Zhang, X.; Kluge, J.; Dorfmann, L.; Kaplan, D.L.; Leisk, G. Silk-based electrospun tubular scaffolds for tissue-engineered vascular grafts. *J. Biomater. Sci. Polym. Ed.* **2008**, *19*, 653–664.

133. Pektok, E.; Nottelet, B.; Tille, J.-C.; Gurny, R.; Kalangos, A.; Moeller, M.; Walpoth, B.H. Degradation and healing characteristics of small-diameter poly(ε-Caprolactone) vascular grafts in the rat systemic arterial circulation.*Circulation* **2008**, *118*, 2563–2570.

134. Zhu, Y.; Cao, Y.; Pan, J.; Liu, Y. Macro-alignment of electrospun fibers for vascular tissue engineering. *J. Biomed. Mater. Res. B* **2010**, *92*, 508–516.

135. Ju, Y.M.; Choi, J.S.; Atala, A.; Yoo, J.J.; Lee, S.J. Bilayered scaffold for engineering cellularized blood vessels.*Biomaterials* **2010**, *31*, 4313–4321.

136. Dutoya, S.; Verna, A.; Lefèbvre, F.; Rabaud, M. Elastin-derived protein coating onto poly(ethylene terephthalate). Technical, microstructural and biological studies. *Biomaterials* **2000**, *21*, 1521–1529.

137. Sell, S.A.; McClure, M.J.; Barnes, C.P.; Knapp, D.C.; Walpoth, B.H.; Simpson, D.G.; Bowlin, G.L. Electrospun polydioxanone-elastin blends: potential for bioresorbable vascular grafts. *Biomed Mater.* **2006**, *1*, 72–80.

138. Wolfe, P.S.; Madurantakam, P.; Garg, K.; Sell, S.A.; Beckman, M.J.; Bowlin, G.L. Evaluation of thrombogenic potential of electrospun bioresorbable vascular graft materials: Acute monocyte tissue factor expression. *J. Biomed. Mater. Res. A***2010**, *92*, 1321–1328.

139. Liu, Y.; Xiang, K.; Chen, H.; Li, Y.; Hu, Q. Composite vascular repair grafts via micro-imprinting and electrospinning.*AIP Adv.* **2015**, *5*.

CHAPTER 5

Assessment of Firefighters' needs for personal protective equipment

Huiju Park1*, Juyeon Park2 , Shu-Hwa Lin3 and Lynn M Boorady4

1 Department of Fiber Science and Apparel Design, Cornell University, 131 Human Ecology Building, 37 Forest Home Dr., Ithaca, NY 14853, USA.

2 Department of Design and Merchandising, Colorado State University, 314A Gifford Building, 1574 Campus Delivery, Fort Collins, CO 80523, USA.

3 Department of Family & Consumer Sciences, Fashion Design & Merchandising, University of Hawai'i at Mānoa, 2515 Campus Rd., Miller 201, Honolulu, HI 96822, USA.

4 Fashion and Textile Technology Department, State University of New York-Buffalo State, TECH 303, 1300 Elmwood Avenue, Buffalo, NY 14222, USA.

ABSTRACT

This study performed focus group interviews with 54 firefighters in four states in the United States to assess the needs for design of firefighters' personal protective equipment. Firefighters' responses indicate three major issues to be considered for improved mobility, comfort and safety: 1) consideration of human factors, 2) sizing and fit, and 3) integrity of protection in the interface between protective equipment and turnout ensemble. Ergonomic design issues were identified through firefighters' responses about their limited mobility of the head and arms while wearing helmet and self-contained breathing apparatus (SCBA), restricted access to coat pockets, back pain and soreness while wearing SCBA. Sizing and fit issues such as the excessive length and bulkiness of glove fingers were identified as a major concern regarding fire gloves resulting in limited mobility and dexterity, negatively effecting firefighters' work efficiency and safety. The integrity of protection in the interface between turnout ensemble, gloves and boots was reported as an important consideration for firefighter comfort, mobility and protection against heat hazards. Possible design solutions were also discussed.

KEYWORDS

Firefighters, Protection Equipment, Wearability Comfort, Mobility, Size Fit

1. INTRODUCTION

Firefighting is one of the most dangerous occupations that require intensive physical work in hazardous environment (Coca et al. 2010). US Firefighters are required to wear turnout ensemble (i.e., protective coat and pants), and other personal protective equipment (PPE) such as helmet, gloves, boots and a self-contained breathing apparatus (SCBA) certified by NFPA (National Fire Protection Association) 1971 and 1981 standards (Figure 1). These items of firefighters' protective system have been designed to provide protection against multiple hazards such as thermal threats (e.g. exposure to flame and excessive heat), toxic gas inhalation, and physical injuries (e.g. cuts, collisions, punctures, slips, falls etc.). So far, the focus of most studies on the firefighters' protective system has been on thermal protection, which combined with advances in material technology during the past decade, has greatly decreased burn injuries (Boorady et al. 2013). However, literature reports that the efforts to improve thermal protection have inevitably increased the weight and bulkiness of the firefighters' protective system, hence, significantly compromising firefighters' mobility and comfort (Adams & Keyserling 1993; Boorady et al. 2013). Sobeih et al. (2006) and Dorman (2007) determined that wearing heavy and bulky turnout clothing caused restricted body movement, leading to a number of firefighters' injuries on unfavorable fire grounds. Coca et al. (2010) also identified a decrease in range of motion at the neck and ankle while wearing turnout ensemble. Moreover, SCBA and boots have also been reported as a cause to impair body balance and musculoskeletal injuries. Carrying a SCBA (weighing 9-13 kg on average) on their back disturbs firefighters' body balance during motion by changing their center of mass, which is a major contributor to fall injuries (Helneman et al. 1989). Additionally, Park et al. (2010) reported that an increase in weight of SCBA elevates the risk of fall and slip injuries. Wearing heavy protective boots, which can weigh up to 4.4 kg, was known to trigger rapid fatigue as evidenced by approximately nine times greater metabolic rate per unit mass, compared to SCBA (Taylor et al. 2011). A previous study by Neeves et al. (1989) also concluded that wearing heavier boots causes more physical burden due to less efficient leg movement during firefighter's job-related tasks.

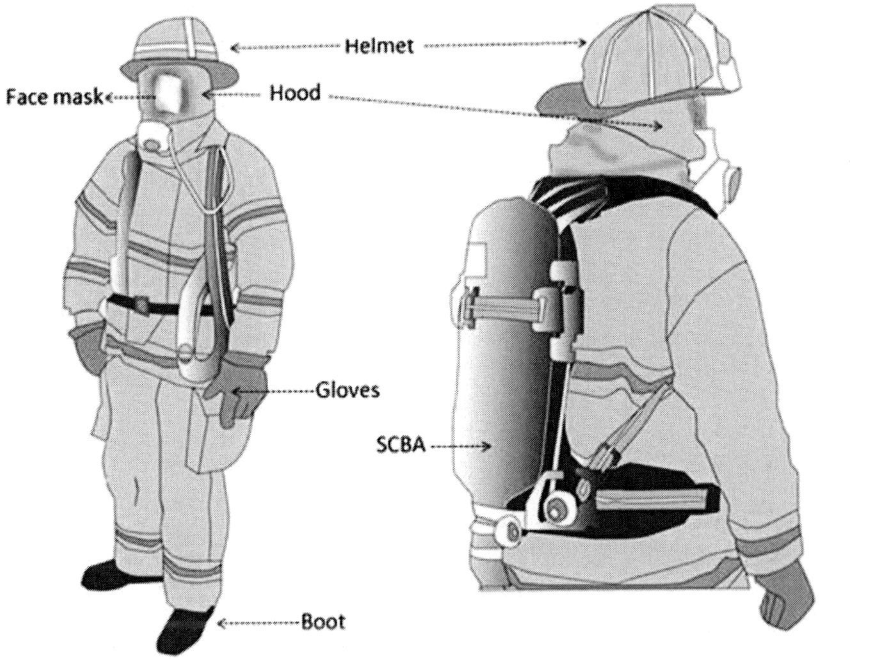

Figure 1 Configuration of firefighters' PPE.

While the aforementioned physiological and biomechanical investigations examine the impact of firefighters' protective system on their mobility and work efficiency providing quantitative evidence, investigation about firefighters' input and needs for protective systems is also an important element in improving the design of their protective system. (Huck 1988) and Akbar-Khanzadeh et al. (1995) claimed that increased wearability and proper usage of the protective clothing system can be achieved only when human factors and user input are included in the design process. The two studies further stated that lack of an understanding of user perception and demand can result in improper or no use of the protective system. Boorady et al. (2013) and Sinden et al. (2013) claimed that firefighters' perception about the performance of their turnout clothing should be considered to improve the design of their protective system. The two studies suggested that designers should take into account the wearers' gender specific issues (e.g., body shape and size) and human factors in their job-related tasks and also upheld that female firefighters are more vulnerable to ill-fitting protective clothing.

Although a few studies investigated firefighters' perception about their turnout clothing, there is no study that investigated their equipment such as helmet, gloves, boots, and SCBA. Considering that the equipment is a critical component of firefighters' protection for their extremities and their head, as well as an essential interface between their body and their job environment, it is imperative to understand user experiences and needs for further design improvements. Moreover, a recent survey of 951 firefighters conducted by National Institute for Occupational Safety and Health reported firefighters' perceived impact of protective equipment on their work efficiency and comfort (Hsiao et al. 2014). Based on the research gap and the actual need for improvement, this study assessed firefighters' perception of their needs for protective equipment (gloves, boots, SCBA and helmet) through a survey and focus group interview (FGI). FGI has been used in the area of social science because it provides an in-depth understanding of research questions by acquiring substantial data from participants (Esterberg 2002). FGI allows participants to speak in depth about the given topics, while they could articulate their perception and ideas independently (Boorady et al. 2013).

2. METHODS

Semi-structured FGIs were performed at fire stations in four different geographical locations in the Unites States. A total of six fire stations participated in this study. The settings of the geographical locations varied from urban to rural and from the Northeastern region to the Western Pacific region. Approval of the Institutional Review Board (IRB) was obtained through the participating universities' IRB offices. Prior to the focus group interviews, a short survey was administered to firefighters. The survey inquired 1) the participant's demographic information (e.g., age and gender), 2) type of commitment to the fire service and duration of the service, and 3) experience of injuries. The survey also included preliminary questions about firefighters' perception of boot and glove performance, considering the significant impact of those two items in extremity protection. These questions were developed by the researchers based on the review of NFPA standards combined with the literature review of firegear studies. The questions concerning the boots inquired the participants' satisfaction relative to 12 properties of the boots, using 5 point Likert-type scale (1: very dissatisfied - 3: neutral-5: very satisfied). The 12 properties included fire/flame protection, hot splash/steam protection, water-proofness, puncture/impact protection, durability, chemical protection, thermal comfort (dryness and comfortable temperature), cushioning, traction quality, structural foot support, stability and flexibility in motion. The questions related to glove performance inquired about firefighters' satisfaction with fire/flame protection, water-proofness, thermal comfort, grip, and dexterity. The entire FGIs were audio-taped for transcription and data analysis. A total of 23 survey questions were asked as summarized in Table 1.

Table 1. Survey questions

Items	Questions
Partic ipants' demographic information	Age
	Gender
Type of commitment to fire service	Commitment type (volunteer vs professional)
	Year of fire service
Injury	Types of injury
	Frequency of injury
Satisfaction level with fire boots	12 key performance properties
Satisfaction level with fire gloves	5 key performance properties

The questionnaire for the FGI consisted of six items (Table 2), asking firefighters' perception about each item of fire gear (boots, gloves, helmet and SCBA) and potential needs for improvement. The researchers followed up with additional questions when the participants' responses required further information or clarification.

A total of 54 firefighters (48 male firefighters and 6 females, age: 34.8 ± 11.1 years), recruited from the four different U.S. locations, participated in the FGI (Table 3). Firefighter participants had 9.8 ± 8.5 years of experience of fire service and detail is shown as follows: 5 participants (<3 years), 10 participants (3-6 years), 6 participants (7-9 years), and 10 participants (more than 10 years).

Table 2. Questions for focus group interviews

Item	Question
Experience	Explain your experience wearing [the name of each item of protective equipment].
Impact	What do you think about the impact of wearing [the name of each item of protective equipment] on your protection, work efficiency, mobility, comfort and safety?
Satisfaction	How satisfied are you with the size and fit of your [the name of each item of protective equipment]? If you are not satisfied, explain why in detail.
Interface	What do you feel about the interface between [the name of each item of protective equipment] and the turnout ensemble (or other items close to/connected with the item of interest)?
Size selection	How do you find the right size?
Suggestion	Do you have any constructive feedback for improvement of the design?

Table 3. Focus group interview participants' information

Site	Professional/Volunteer	Geographical region	Total # of participants
1	2/7	Northeast	9
2	10/0	West	10
3	10/0	West	10
4	11/0	Pacific	11
5	6/0	Pacific	6
6	0/8	Northeast	8

3. DATA ANALYSIS AND RESULTS

Descriptive statistics of participants' satisfaction with their fire gear and frequency of injuries were calculated to identify the potential area of improvement in performance of the fire gear. Firefighters' comments from the FGI were coded using the interpretive thematic analysis method, which allowed grouping of the responses into overarching themes with related issues.

4. SURVEY

4.1 Fire gloves

The mean value of firefighters' responses about the performance of fire gloves revealed low satisfaction with water-proofness (mean: 1.9 ± 1.1), grip (mean: 2.5 ± 0.9), dexterity (mean: 1.8 ± 0.9) and thermal comfort (mean: 2.9 ± 0.7). However, firefighters' satisfaction with fire/flame protection was relatively higher (mean: 3.6 ± 1.03) (Figure 2).

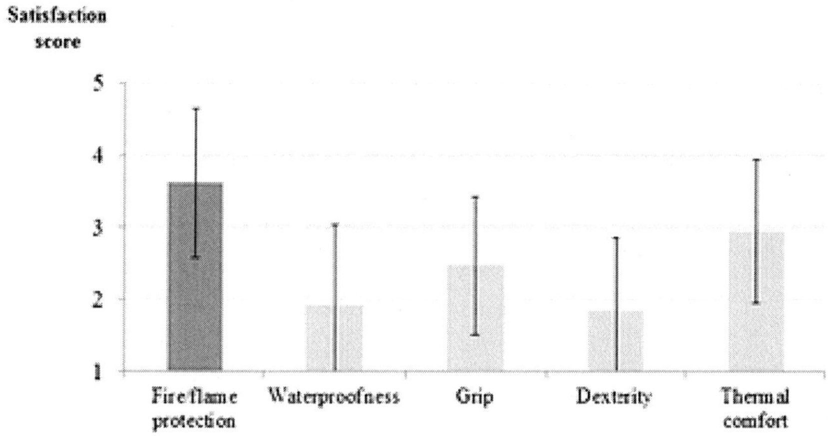

Figure 2. Mean values of firefighters' satisfaction with 5 properties of fire gloves.

4.2 Turnout boots

Firefighters' responses about satisfaction with cushioning (mean: 2.8 ± 1.2), structural foot support (mean: 2.5 ± 1.3) and flexibility in motion (mean: 3.0 ± 1.2) indicated low mean values (≤3: neutral), while the other nine properties regarding the levels of protection, durability and thermal comfort showed relatively higher satisfaction scores (Figure 3). The highest satisfaction levels were associated with water-proofness (mean: 4.4 ± 0.7) and fire/flame protection (mean: 4.2 ± 0.7).

4.3 Reported injuries

Many firefighters indicated wearing ill-fitting boots especially bulky rubber boots caused blisters (n = 21) on the calves, around the boot collar and plantar foot (Table 4).

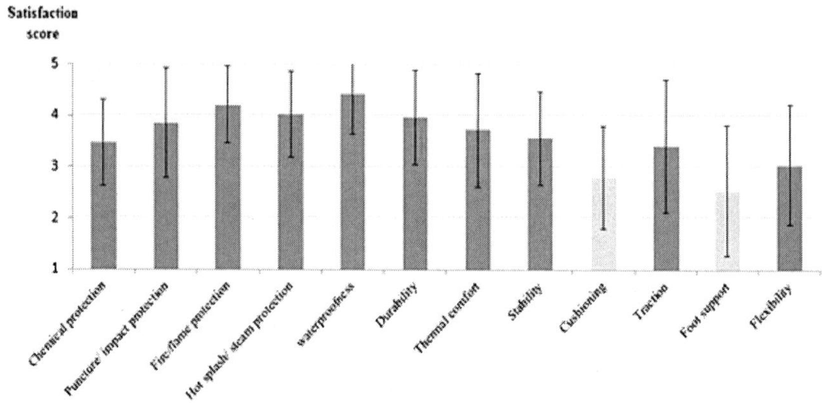

Figure 3. Mean values of firefighters' satisfaction with 12 properties of fire boots.

Table 4. Reported injuries

	Number of incidences			
Injuries	**1-5 times**	**6-10 times**	**Over 10 times**	**Sub-total**
Blisters	13	4	4	21
Muscle pain	9		11	20
Stumbles	8	1	9	18
Slips	8		7	15
Falls	10		3	13
Ankle sprains	10			10
Bruises	3			3
Minor cuts	1			1
Smashing	1			1
Scrapes	1			1
Burns	1			1
Electrically conducted	1			1

Muscle pains (n = 20) in the back, shoulders, back of the neck and knee were also reported as frequent injuries. This may be related to the fact that firefighters carry heavy SCBA or tools while working in bulky turnout ensemble and equipment that may restrict or alter their normal body movement. More than ten firefighters reported they experienced stumbles, slips, falls, and ankle sprains as they worked on unfavorable fire grounds or confined spaces (Table 4).

5. FGI

5.1 Fire gloves

Firefighters reported poor dexterity as the biggest concern with gloves, and it was due to oversized glove design with multi-layered bulky structure. They reported that the poor dexterity and limited grip hindered them from working with tools. For example, a female firefighter accounted as such: *"Dexterity is the big issue, especially if they're smaller hands. Like for me, my hands are really small so it's hard to find gloves that fit really good (Site #2)."* Male firefighters also indicated the dexterity issue associated with gloves:*"The fit, the way they form the fingers they are just too fat, there's too much material where is the ability of protection and the ability to grip right, there's just a lot of extra bulk on the fingertips (Site #1)."* Another male firefighter described his experience with gloves in a specific firefighting task: *"We have to tie a knot. If you're working with extra this much on the end of each finger, trying to dexterously feed up a rope through a knot, it's very difficult (Site #6)."* Most firefighters expressed the stiffness of material had made gloves hard to crap and pick small devices. Very often, firefighters cannot even operate a phone to communicate each other due to thickness and stiffness materials of gloves.

Challenging tasks with gloved hands include a majority of essential movements requiring hand motions such as using a walkie-talkie, carrying, holding and using tools or hoses etc. In particular, holding a hose (approximately 45 kg of pressure at the tip) with gloved hands is very challenging due to poor grip in the palm and fingers. A male firefighter elucidated the situation as such: *"There's almost no grip on my palm right now. Even though it's resting on my palm, all the grip is in the fingers. I can't grip the hose with my gloves. I grip it with my body and my arm. If I try to hold it just with my glove, it will slide (Site #6)."*

The challenge resulted from oversized gloves seemed more obvious to small-handed firefighters, especially females, prompting them to use better-fitting work gloves (lacking fire and flame protection) for non-fire related duties. Firefighters also reported that donning the wet gloves once they had been taken off was difficult because the gloved fingers tended to turn inside out when their hands were wet, thus making it hard to rearrange them. This seemed to render firefighters hesitant to take off their wet gloves despite thermal discomfort and loss of grip. Such firefighters' behavior in the field could cause increased risk of burn injury from quicker heat transfer in the wet gloves compared to dry gloves, as described in the following account: "*My hands sweat from doing stuff with water and then I pull the glove off and the whole lining comes off with my hand. And I can't get it back in", "If you were holding a hose and you get a lot of splash back on your gloves, you gloves can get wet and be very hot right away (Site #6).*" Another account also illustrated the firefighting risk factor associated with wet gloves: "*Due to the poor dexterity, firefighters often feel temptation to take gloves off. For example, when you need to use axes and your gloves get wet, think about it.... Using an ax and wearing slippery gloves is not a good match (Site #6).*" Additionally, a wide range of glove sizes offered by manufacturers were also reported as a confusing factor in finding the right size, as a firefighter asserted like: "*You get a box of gloves and you just try them on until you find one (Site #6).*"

A few female firefighters indicated that working with heavy tools while wearing oversized gloves imposes more physical challenges to them. "*Holding it (the chainsaw) up is such a pain in the butt because your fingers are so thick in the gloves. Starting that chainsaw is very hard for me. I have to literally sit on the chainsaw and pull like this to get it started (Site #6).*"

5.2 Turnout boots

Firefighters reported that wearing turnout boots changes their walking pattern and the extent of the change is greater for rubber boots than leather boots. Firefighters described their gait, while wearing turnout boots like: "*You have to have extra kick in your boot.*" "*You can't really run (Site #6).*" In particular, firefighters reported various reasons for discomfort when wearing rubber boots, including poor fit, low resistance against abrasion (Figure 4), heaviness, bulkiness, stiffness and difficulty in finding the right size.

Furthermore, firefighters reported that poor fit is a major reason for changes in their walking patterns and this negatively impacts on their safety on the fire ground. They also indicated that they experienced frequent fall-offs of the rubber boots due to the poor fit, while ascending or descending during ladder operations, stair climbing, and getting on/off the fire engine. Firefighters additionally accounted their frequent experiences of blisters due to the increased rubbing between their feet and boots resulting from the poor fit. Moreover, they reported that the stiffness of rubber often caused chaffing injuries to their calves, as a female firefighter described as following: "*I had two pairs of thick socks and I always had blisters in the rubber ones (Site #1).*" Additionally, firefighters reported that wearing rubber boots during winter leaves their feet vulnerable to cold injuries and likely causes loss of traction on the slippery fire ground: "*They (rubber) didn't keep your feet warm, I remember being at a fire in the winter –my feet were frozen (Site #1).*" "*When it's really cold, the rubber boot freezes up so you don't have good traction (Site # 2).*"

Figure 4. Worn out toe cap due to frequent friction on the ground while crawling.

Firefighters showed a strong preference for leather boots due to the following reasons: a snugger fit, lighter weight, better flexibility, ankle support, tactile sensation on the ground, more comfortable foot motion and easier size choice than rubber boots. *"I find it's better flexibility when your foot is bent in the leather boots, you couldn't get that flexing. Where you are planting your foot of pushing off, the more pliable the material the better it's going to be. (Site #1)," "When I got my leather boots I felt like I didn't have to be as careful as I moved my leg because my boots weren't going to fly off anymore (Site #6)."*Some firefighters also described their positive experiences with leather boots, in comparison with rubber boots: *"I've used rubber boots for years, and what I find with the leather ones is you have better tactile sensation when you're climbing a ladder you can feel the ladder better (Site #3)."*

5.3 SCBA

Firefighters predominantly indicated their experiences of discomfort wearing SCBA, because of restricted movement of the head and the arm, soreness on the shoulder, and back pain. Specifically, they often experienced restricted motion of the head, especially while looking up and around because the back brim of their helmet hits the top portion of the air pack (Figure 5). They indicated that wearing SCBA had a greater impact on firefighters whose height is relatively short, because they carry the same air pack but have a shorter torso than their taller colleagues; Shorter firefighters indicated that they could not place the SCBA further down on the back, as did their taller colleagues to prevent the back brim of the helmet from hitting the top portion of the SCBA for improved neck mobility.

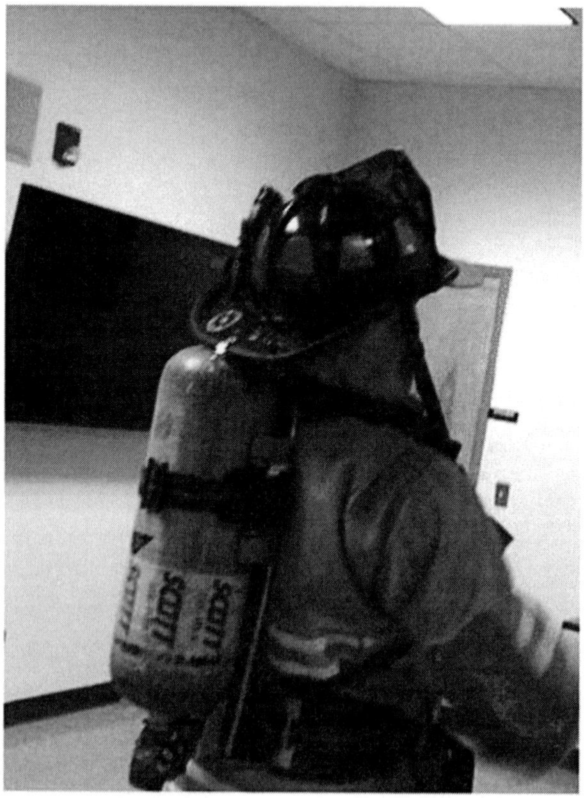

Figure 5. Limited mobility due to physical conflicts between the helmet and air pack.

The following accounts explained how their physical profile, especially the torso size, affects neck mobility and visibility of firefighters in the field, in relation to SCBA: *"You know a lot of time it restricts your ability to look up. So, just depending on how tall you are and where your air pack comes to, on your back. Looking up either when you're walking or crawling can be an issue (Site #2).", "I find a lot of restriction in the shoulders- overhead work especially with the SCBA (Site #1)."* In addition, some firefighters pointed out that the stiff and oversized neckline of the turnout coat is an additional reason for their limited neck mobility. Firefighters mentioned that wearing SCBA restricts their arm movement considerably: *"You can't reach up – your jacket sort of moves with you but if you've got the SCBA buckled around you (Site #1)." "Sometimes the straps do limit your mobility and your access to different things like when you have to reach over head (Site #6)."* Another common response was a hindrance in accessing the pockets because the SCBA straps tend to block access to pockets which likely impedes firefighters from reaching tools in the pockets when they need to use them: *"I don't think a lot of people put the stuff in the coat pockets because it gets tangled up with your SCBA pack. Most people just use their leg pockets (Site #3)."*

Firefighters also mentioned concerns about possible injuries when they fall or slip due to the heavy SCBA on their back, as illustrated in the account: "*If you fall on your back there's potential to have significant injury. If you were to fall, slips on the ice and you fall flat on your back, you're probably going to hurt yourself (Site #2)."*

5.4 Helmet, face mask and hood

Firefighters reported discomfort from ill-fitting face masks when they were worn with the helmet as such: *"When I have my mask on, I can't push my helmet down where its comfortable so I feel like it's sitting up higher on my head and I have to crank my chin strap down to keep it from falling off and it just doesn't marry up well on my face, and it's probably because my face is smaller (Site #2)."* Some firefighters also reported the itchiness of the hood, especially when it is wet.

5.5 Interface between equipment and turnout ensemble

Firefighters reported the interface between equipment and turnout ensemble as an area of improvement for enhanced protection and mobility. Although each individual item of PPE may be well designed to meet NFPA standards, combining multiple pieces of equipment together creates a bulky turnout ensemble. Firefighters also need to wear their turnout ensemble on top of daily clothing, and it appears to add additional bulkiness and inconvenience to the clothing interface, especially where gloves and boots are connected with the turnout ensemble. This situation causes mobility restrictions, discomfort, poor-fitting and body exposure to heat hazards during motion, as the following account states: *"They're designed individually and individually they are great, but together all these layers stack up and they're not designed to work together necessarily. If somebody could have some continuity of thought across each layer, which would help bulking up in the kind of bulky areas (Site #2)."*

Evidently, firefighters are concerned about the exposure of their lower limbs to heat hazards while they are crawling, as they frequently experience their pants riding up above the boots, exposing their calves to heat hazard: *"When we're crawling that boot design is cut a little bit lower so when we'd crawl, our pants would ride up, so then they got exposed areas and you know there's nothing to keep that from happening (Site #2)." "When I would be doing a search my pants would ride up over my boots and expose skin and then when I would stand up to go from a crawling to a standing position my pants would hang up on the tops of my boots (Site #3)."* Excessive bulkiness in the interface between the sleeve and gloves was also reported as an area in need of improvement: *"They're really thick cuffs so nobody sticks them underneath because then you've got more mobility issues of another layer of material (Site #2)."* Major issues that were discussed at the FGI were summarized in Table 5.

6. DISCUSSION AND CONCLUSIONS

Input from firefighters in this study, related to their PPE, identified several areas of improvement. The three major issues that firefighters' response indicated are a) consideration of human factors, b) sizing and fit, and c) integrity of protection in the interface for improved mobility, protection and comfort.

Ergonomic design issues were identified through firefighters' responses about their limited mobility of the head and arms while wearing helmet and SCBA, restricted access to coat pockets, back pain and soreness while wearing SCBA. A study by Coca et al. (2010), which measured range of motion of firefighters while wearing a turnout ensemble, has also reported limited mobility around the neck area. Since this limited range of motion leads to limited vision, improving the range of motion of the neck, changing the helmet design, possibly with a smaller back brim or an entire helmet, or changing the SCBA design, will have potential to improve firefighters' safety. This will further enhance their ability to visually inspect and recognize risk factors in the field where there are a variety of hazards, especially due to damaged building structures.

Table 5. Summary of major considerations for improved design

Items	Issues to consider improvement of design
Fire gloves	Needs to improve dexterity
	Oversized glove design
	Lack of grips
	Confusing sizing system and inconsistency in size among manufacturers
	Potential heat hazards when gloves get wet
Turnout boots	Change in walking pattern in turnout boots
	Frequent fall-offs due to poor fit
	Coldness and lack of traction of rubber boots during winter
	Strong preference for leather boots (snug fit, better tactile sensation & lightweight)
SCBA	Restricted movement of the head and arms
	Greater negative on shorter firefighters
	muscle soreness and pain in the back and shoulder
	Hindrance in accessing pockets of turnout coat
	Risk of serious injury when falling
Helmet, face mask and hood	Poor fit
	Itching face mask when wet
Interface between equipment and turnout ensemble	Excessive bulkiness where the boots and gloves are connected with turnout ensemble
	Exposure of calf to heat hazard resulting from the pants riding up

Firefighters' input in this study also showed that they experience soreness in the shoulder and back, which is related to wearing SCBA. Previous studies found that wearing SCBA is positively related to loss of body balance and fall injuries on the fire ground by shifting the body's center of gravity and causing gait instability (Park et al. 2010). By wearing the SCBA (weighing about 30–40 pounds) on the back, the firefighter's center of gravity shifts upward and backward from the pelvis (original location of unloaded body). The gravity shift caused by wearing the heavy SCBA may have greater impacts on mobility to lighter-weight firefighters, since the ratio of body mass to weight of load carriage is a critical factor to fatigue increase (Marshall 1980). Similarly, the negative impact of wearing SCBA can be greater to shorter firefighters who may have less body mass as the shift of center of gravity is dominated by the additional weight and the dimensions of the air pack, relative to the body profile; taller firefighters have a relatively longer torso allowing them to lower the SCBA resulting in improved neck mobility and possibly alleviating the impact of SCBA on the shifting of the center of mass. Griefahn et al. (2003) demonstrated that lowering the weight distribution of the air pack on the back could improve firefighters' mobility and work experience. They reported that rucksack style SCBA, featuring a lower placement of the air pack and an even weight distribution throughout the back, resulted in faster task completion with less fatigue than a conventional cylindrical shaped SCBA. Based on the aforementioned ergonomic consideration and previous research, an attempt to lower the placement of the air pack may minimize the shifting of the center of gravity, as well as avert physical conflicts with the back brim of the helmet, thus improving body balance and the range of motion around the neck area in firefighters. Research and design efforts to distribute the weight of the air pack more evenly around the torso are necessary in order to alleviate back pains and muscle soreness. Park et al. (2013) study affirmed that wearing a backpack can result in chronic lumbar pain over time. Furthermore, relocating coat pockets should also be considered in order to enhance firefighter access to their tools while wearing the SCBA.

Sizing and fit issues were identified as a major concern regarding fire gloves and boots, resulting in limited mobility and dexterity, negatively effecting firefighters' work efficiency and safety. Firefighters clearly pointed out that oversized gloves significantly limit grip and dexterity, resulting in additional physical effort requirements in a variety of tasks. In our study, a commonly reported issue was the excessive length of glove fingers. These over-sized gloves are a more serious issue for female firefighters compared to their male colleagues. Female firefighters tended to feel that the majority of PPE, including gloves, is designed based on male physique. Female firefighters reported that the loss of grip and dexterity from wearing oversized gloves is an extra challenge for them, and it added more stress to their relatively weak physical strength. This leads female firefighters to develop their own strategies to perform tasks that require using tools with oversized gloves, in order to compensate their limited physical strength, resulting in supplementary physical strain (Sinden et al. 2013). Firefighters also reported that a wide range of inconsistency in glove sizes by different manufacturers is another challenge when choosing the right size of gloves. A similar sizing and fit issue was reported regarding rubber boots in this study. Firefighters mentioned that rubber boots are oversized, leading to frequent fall off of the boots during ascending and descending movements, and also blisters due to poor fitting and stiff boot materials. This may also be related to the firefighters' frequent experience of fall injuries, ankle sprains, stumbles, slips, and blisters reported in the survey of this study. Sizing and fitting issues were also identified in respect to the face mask and helmet. The aforementioned sizing and fit issues in firefighters' PPE suggested needs for revisiting anthropometric data and re-evaluating the sizing system of PPE.

The integrity of protection in the interface between turnout ensemble and PPE (gloves and boots) was reported as an important consideration for firefighter wearing comfort and protection against heat hazards. Reducing bulkiness in the sleeve-gloves interface needs to be considered for improved comfort and wearability. Firefighters' concerns about exposing the calves due to pants' riding up above the boots while crawling indicated a need for additional research and design effort to maintain the integrity of thermal protection in the interface between the turnout ensemble and extremity protection with consideration of the interaction between the moving human body and PPE.

This study performed FGI with 54 firefighters (including only six female firefighters) and was limited to four States in the United States. Therefore, generalizing the findings of this study for the entire fire service population requires some caution. Nevertheless, the study presented practical implications and insight, both for the fire-gear industry and apparel design educators, based on user input. Future studies with a larger sample and firefighters' performance evaluation in a simulated field environment could provide additional meaningful information.

ACKNOWLEDGEMENT

This research was supported in part by the Cornell University (Project No. NYC 329832), Colorado State University (COL00224), and University of Hawai'i at Mānoa (HAW00360-R), Agricultural Experiment Station funds, received from the National Institute of Food and Agriculture (NIFA) United States Department of Agriculture. Any opinions, findings, conclusions, or recommendations expressed in the publication are those of the author(s) and do not necessarily reflect the view of the National Institute of Food and Agriculture (NIFA) or the United States Department of Agriculture (USDA).

COMPETING INTEREST

All authors declare that they have no competing interest in relation to this study.

AUTHORS' CONTRIBUTIONS

HP and JP collaboratively designed the research protocol and developed research instruments and a coding scheme for data analysis. All authors carried out focus group interviews and coded data. HP drafted the manuscript and JP, S-HL and LB participated in the revision process. All authors read and approved the final manuscript.

REFERENCES

1. Adams, PS, & Keyserling, WM. (1993). Three methods for measuring range of motion while wearing protective clothing: a comparative study. *International Journal of Industrial Ergonomics, 12*(3), 177–191.

2. Akbar-Khanzadeh, F, Bisesi, MS, & Rivas, RD. (1995). Comfort of personal protective equipment. *Applied Ergonomics, 26*(3), 195–198.

3. Boorady, LM, Barker, J, Lee, Y-A, Lin, S-H, Cho, E, & Ashdon, SP. (2013). Exploration of firefighter turnout gear; Part 1: identifying male firefighter use needs. *Journal of Textile and Apparel, Technology and Management, 8*(1), 1–13.

4. Coca, A, Williams, WJ, Roberge, RJ, & Powell, JB. (2010). Effects of fire fighter protective ensembles on mobility and performance. *Applied ergonomics, 41*(4), 636–641.

5. Dorman, LE. (2007). *PhD Thesis, The Effects of Protective Clothing and its Properties on Energy Consumption During Different Activities.*: © Lucy Ellen Dorman.

6. Esterberg, KG. (2002). *Qualitative Methods in Social Research*. Boston: McGraw-Hill.

7. Griefahn, B, Künemund, C, & Bröde, P. (2003). Evaluation of performance and load in simulated rescue tasks for a novel design SCBA: effect of weight, volume and weight distribution. *Applied ergonomics, 34*(2), 157–165.

8. Hsiao, H, Whitestone, J, Kau, TY, Whisler, R, Routley, JG, & Wilbur, M. (2014). Sizing Firefighters method and implications. *Human Factors: The Journal of the Human Factors and Ergonomics Society.*

9. Helneman, EF, Shy, CM, & Checkoway, H. (1989). Injuries on the fireground: risk factors for traumatic injuries among professional fire fighters. *American Journal of Industrial Medicine, 15*(3), 267–282.

10. Huck, J. (1988). Protective clothing systems: a technique for evaluating restriction of wearer mobility. *Applied ergonomics, 19*(3), 185–190.

11. Marshall, SLA. (1980). *The Soldier's Load and the Mobility of a Nation.* Quantico, Virginia: Marine Corps Association.

12. Neeves, R, Barlow, DA, Richards, JG, Provost-Craig, M, & Castagno, P. (1989).*Physiological and Biomechanical Changes in Fire Fighters Due to Boot Design Modifications. International Association of Fire Fighters and the Federal Emergency Management Agency.*

13. Park, H, Branson, D, Petrova, A, Peksoz, S, Goad, C, Warren, A, Jacobson, B, & Kamenidis, P. (2013). Effects of body armor and load carriage on lower limb joint movement. *Journal of Human Performance in Extreme Environments, 10*(2), 3. doi:http://dx.doi.org/10.7771/2327-2937.1049.

14. Park, K, Hur, P, Rosengren, KS, Horn, GP, & Hsiao-Wecksler, ET. (2010). Effect of load carriage on gait due to firefighting air bottle configuration. *Ergonomics, 53*(7), 882–891.

15. Sinden, K, MacDermid, J, Buckman, S, Davis, B, Matthews, T, & Viola, C. (2013). A qualitative study on the experiences of female firefighters. *Work: A Journal of Prevention, Assessment and Rehabilitation, 45*(1), 97–105.

16. Sobeih, TM, Davis, KG, Succop, PA, Jetter, WA, & Bhattacharya, A. (2006). Postural balance changes in on-duty firefighters: effect of gear and long work shifts. *Journal of Occupational and Environmental Medicine, 48*(1), 68–75.

17. Taylor, NAS, Lewis, MC, Notley, SR, & Peoples, GE. (2011). The Oxygen Cost of Wearing Firefighters' Personal Protective Equipment: Ralph Was Right! In S Kounalakis & M Koskolou (Eds.), *ICEE 2011 XIV International Conference on Environmental Ergonomics: Book of Abstracts* (pp. 236–239). Athens: National and Kapodestrian University of Athens.

CHAPTER 6

Systems thinking in designing automotive textiles

Pammi Sinha*, Subramanian Senthilkannan Muthu, Iain Taylor, Rita Schulze, Katharine Beverley, Claire Day and Nicoleta Tipi

Sinha, Beverley, Day, Schultz: Department of Fashion and Textiles, University of Huddersfield, Queensgate, Huddersfield HD1 3DH, UK

ABSTRACT

We present the complexities in terms of designing automotive exterior seating materials (seat coverings and interior linings) at Sage Automotive Interiors (UK), which is a division of a global international automotive textile supplier with headquarters in the US. Sustainability and innovation are emphasized in documents communicating the company's vision. Using a case study approach, we consider the current design, development and manufacture process and examine it for the potential for feedback loops, identify potential leverage points to effect change and how the process could divert wastes from disposal. We will highlight where sustainable decisions can be incorporated and the difficulties in achieving true sustainability. We argue that a systems approach is needed from conception to final product to ensure economic recycling of textiles and fibres used in automotive seating. Without which, the reality is at best incineration for energy and/or landfill, thus losing important, finite resources forever from a diminishing resource pool of raw materials.

KEYWORDS

Automotive textiles, Sustainability, Systems, Recycling, Incineration, Waste, Landfill

1. BACKGROUND

1.1 Systems thinking

Systems may be defined as a set of independent but interlinked phenomena; identifying and using beneficial relationships and linkages between different parts of a system is key to optimizing the whole (Charnley et al. 2011; Mingers and White 2010). Meadows (1997) identified nine places of intervention to influence and change systems: (9) numbers - subsidies, taxes and standards, (8) material stocks and flows - substituting with more sustainable material, (7) regulating negative feedback loops - control undesirable factors by legislating to minimize harmful materials or processes, e.g. REACH, (6) driving positive feedback loops - encourage desirable factors through promotion and aid for sustainable practices, (5) information flows - working towards transparency in the supply chain, (4) rules of the system - develop existing or new practices and policies beneficial to all stakeholders, (3) the power of self-organization - internal intervention akin to organizational or self-learning, (2) the goals of the system - profit to survive and compete, but now incorporate social and environmental concerns, and (1) the mindset or paradigm out of which the system arises - the accepted status quo, influencing this can bring about the biggest changes. All systems have inputs, outputs and feedback mechanisms; for this paper, we examine the design process for automotive textiles and the system within which it operates in to outline the implications for environmental product design.

1.2 Sustainability in the automotive industry

The European Union's end-of-life vehicles (ELV) directive stipulates that, by 2015, 95% of new vehicles need to be recyclable (EURAPA 2010) and that 'vehicle manufacturers and material and equipment manufacturers must: (a) endeavour to reduce the use of hazardous substances when designing vehicles, (b) design and produce vehicles which facilitate the dismantling, reuse, recovery and recycling of end-of-life vehicles, (c) increase the use of recycled materials in vehicle manufacture and (d) ensure that components of vehicles placed on the market after 1 July 2003 do not contain mercury, hexavalent chromium, cadmium or lead, except in the applications listed in Annex II of the Directive (Normand 2008). Manufacturers are requesting textiles that disassemble and are recyclable; suppliers are being requested to recall their products after use to recycle them (Normand 2008), potentially forcing readjusting their operations to support this new additional element to their business. Many original equipment manufacturers (OEMs) currently use the design for disassembly processes using compatible chemicals, components and similar fibre types to increase their recycling and reuse efficiencies. The automotive industry is interesting because environmental sustainability of textiles is not (yet) the main focus of the car companies' sustainability efforts, which centre on fuel efficiency and lightweight cars. OEMs are therefore much more hesitant to make improvements regarding EPD for those 'peripheral' parts of the car which customers are less likely to ask about and where the influence on sales and reputation is perceived as uncertain. On the other hand, seating textiles are visible to the customer, and so, it is important to understand the internal and external pressures which the design team faces. Therefore, the design process is analysed to get a detailed understanding of the factors driving the design decisions.

1.3 The automotive assembly system

Often described as tiered, up to four tiers excluding the OEMs, tier 1 supplies interior compartments, e.g. seats, engines, floors; tier 2 carries out subassembly activities such as cut and sew and supply parts such as headliners; tier 3 supplies the manufacturing for the upholstery whilst tier 4 supplies the materials (Shishoo 2008). Tiers 1 and 2 companies may be integrated, and 'cut and sew' operations may be carried out by the same company. Fabric laminators supply their services the cut and sew operations (Tier 2). The fabric formation companies may also conduct fabric lamination in-house or out-source. There are many permutations; companies may supply yarn directly to the fabric formation company, foam and scrim directly to a fabric lamination company or only yarn to the fabric formation company and the fabric formation company supplies the scrim to the fabric lamination company. Moreover, the three companies may be integrated (Taylor 2012). An example is the consolidation by tier 1 companies. The OEMs, frustrated by the increasingly longer lead times and operational constraints, only worked with fabric suppliers with cut and sew operations. Tier 1 companies (e.g. Johnson Controls) have bought out fabric companies and become vertically integrated companies with control of upstream *and* downstream activities making the full range of products from seats, instrument panels, headliners, etc. With 25% of the market, Johnson

Controls vertical integration captures increasing amount of value along the chain but loses flexibility and speed. The independent status of Sage AI is advantageous because it affords flexibility and speed within the system (Taylor2012).

2. METHODS

A single case study, according to Yin (2009), was conducted using a number of semi-structured interviews from November 2011 until July 2014 (on site and off site), with the managing director, design and development director, designers and members of the logistics and technical team. The interviews were recorded, transcribed and triangulated with the company's documents and literature regarding product development process. The analysis was validated by the design and development director. The commercial sensitivity, the detailed nature of the analysis and associated time requirements necessitated a single case study. Sage AI (UK) was chosen as a case study company due to their strong interest in environmental product development and existing close working relationships with the University.

To analyse the design process, we used Integration Definition for Function (IDEF) modelling, a standardized and widely used method to model manufacturing processes (Rashid and Ismail 2007). Used as a visual tool, it helps describe, analyse and communicate ideas (Yu and Wright 1997). It has proven effective for detailed system analysis (Knowledge Based Systems 2010). The modelling often starts from a description of the process at the highest level, which is then broken down hierarchically to the level of detail required for a meaningful analysis. This allows the user to describe the details of complex processes whilst keeping a good overview of the activities (Sugiyama et al. 2008).

IDEF-0 has also been used in the past to support corporate decision-making processes and to support the redesign for the improvement of processes (Hernandez-Matias et al. 2008). In recent years, IDEF-0 has been used to help introduce environmental criteria into business processes. For example, IDEF-0 models were used to illustrate new process designs, considering environmental and health criteria (Hirao et al. 2008) or to show how risk assessment and life cycle assessment can be facilitated using IDEF-0 to help understand the process and information requirements (Kikuchi and Hirao 2009). Kikuchi et al. (2012) used IDEF-0 to illustrate a systematic assessment process for environmental, health and safety aspects in the process development phase. Sugiyama et al. (2008) used IDEF-0 as a modelling tool to support the integration of LCA in process design - the method helped them to highlight information requirements for the newly required activities.

According to the IDEF-0 standard, the purpose and viewpoint, system boundaries and meaning of the arrows should be defined. For this study, the purpose of the model was to understand the design development process of the case study company and to identify where environmentally relevant decisions are made in this process. The model takes the perspective of the design and development director, with the intention to focus on the steps in the design and product development process stages. As a consequence, the model includes all processes which the design team are directly involved in, starting with research and development, design concept development and selection by client, up to the point where a master sample of the fabric has been selected as a template for production. Figure 1 demonstrates a basic structure of the IDEF-0 (National Institute of Standards and Technology1993; Kikuchi et al. 2012). The control arrows in the IDEF-0 model represent factors which need to be taken into account in the respective process step. They are the most relevant factors for the identification of the barriers.

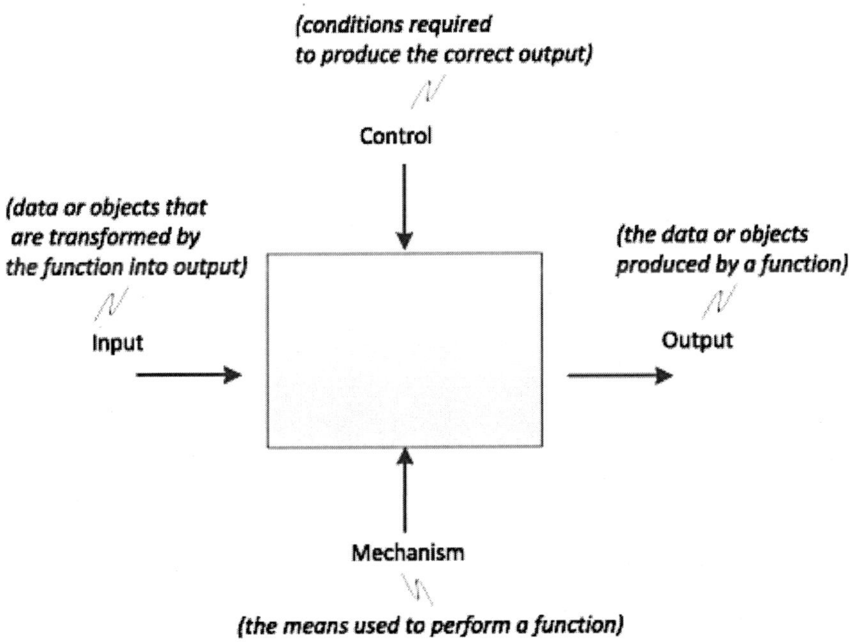

Figure 1 Basic structure of IDEF-0, based on (National Institute of Standards and Technology **1993**).

The purpose of the IDEF-0 modelling was to obtain a detailed examination of the product design/development process at the case study company and was used as a basis to identify the following:

1) where in the design process the most environmentally relevant decisions are being made

2) the crucial factors affecting these decisions, and

3) barriers to and enablers for environmental product development in the current design process, using the introduction of recycled yarn as a particular case study example.

3. RESULTS AND DISCUSSION

3.1 Sage AI (UK) company background

Sage AI, originally a part of the former Milliken Automotive and the Viktor Achter Groups. Sage Automotive Interiors (Sage AI) business is headquartered in the US, and solely in the automotive sector with bases in Poland, Japan, India, South America and China. The Sage AI business is entirely related to the automotive industry. Their fundamental business is design, incorporating engineered and technical qualities to add value to their products. Sustainability and innovation is paramount to their vision.

Sage AI produces textile interiors based on two manufacturing routes: woven or knitted. Most developments are related to enhancing these processes, e.g. through embossing, stitch and HF welding. The samples are developed through their plants overseas; thus, knowledge of their capabilities is essential. This knowledge accrues through experience, and the creative or technical designers visit the partner manufactures every month. Most innovations are for enhancing the fabrics, e.g. rotary embossing, 12 head embroidery machine and HF welding. Various OEMs now produce shorter runs and offer special editions to add more interest to the colour and design of the automotive Powell NB (2006). The use of embroidery is a new technical capability, enabling mass customization service. When developing initial ideas (for either internal or external clients), Sage AI liaise with their suppliers' technicians to establish feasibility with partner companies; where, if it is not possible, they will develop new supplier relationships and run feasibility trials.

Sage AI (UK) own manufacturing plants in Poland (weave) and Egypt (knit). The creative or technical designers visit the manufacturers monthly to maintain communication and address any design/production issues. Costs limit fibre choice usually to polyester which is then designed for variety of looks, e.g. highly lustrous, soft matt, textured and novelty yarns to create interest in the woven and knitted fabrics. Wool can be used but is expensive, therefore limited (Sage used wool for the BMWi3). Yarns are supplied by Autofil Worldwide Ltd., who also piece- or yarn-dye and texturize to a very wide variety of looks and feel; the blend, has to be visually appealing to the customer (the OEM), pass various rigorous tests (such as abrasion and light fastness), be durable (up to 10 years) and most importantly, cost effective.

3.2 Application of textiles to seating

The manufactured polyester fabrics are usually laminated with foam and a lightweight fabric 'scrim'. These are manufactured to a schedule and then shipped to a cut and sew company. These companies, often part of the final seat manufacturer, receive the fabric in roll form and mechanically cut parts which are then sewn together to create a seat cover. The covers are then usually shipped, often over long distances, to where the seat itself is being made, where they are incorporated into the final seat. The seat is then sequenced in trackside as the car they are destined for is being made just in time; the seat assembler receives notification of which type, colour and style manufacture of the seat is required a few hours before it is required to meet the rest of the car on the assembly line. Other parts, e.g., engines, headliners, instrument panels, doors etc. are equally sequenced in.

3.3 The design and product development process

The product development process may take between 3 and 5 years from brief setting to final vehicle production. This is extended by the additional 7 years approximate length of time that the textiles will be in vehicle. Figure 2outlines the process from concept to master sample; for ease of describing the process, stages have been assigned various developmental stages of the process (stages A1 to A4).

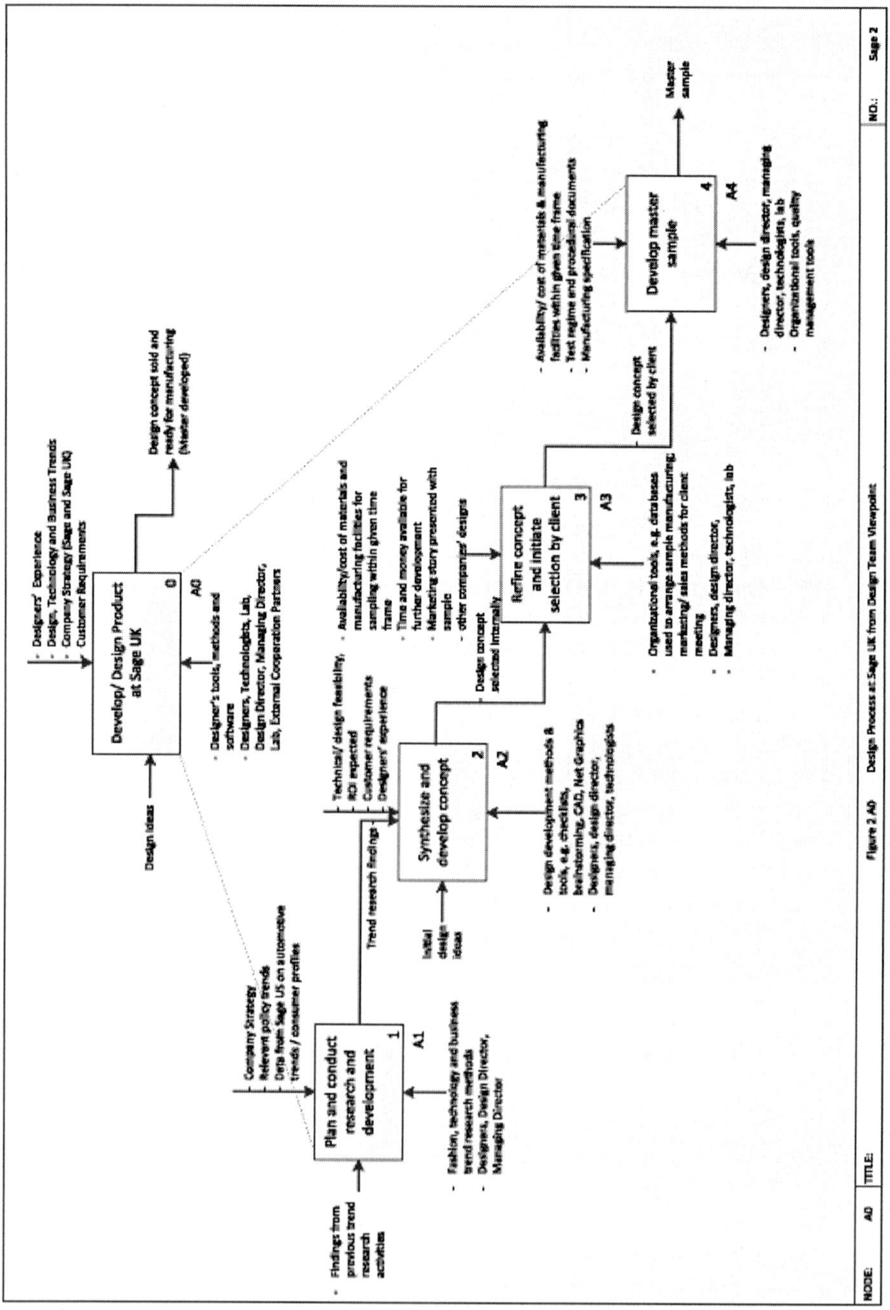

Figure 2. Product development at Sage AI (concept to master sample).

Product development or innovation begins with a brief, either from the customer/client (OEM, which designers interviewed termed as *client based*) or in-house through their own product/consumer knowledge (e.g. designer, manager, etc. - the designers interviewed termed this as *open-line brief*). Sage AI developed FXC for Honda from a brief set by Honda, whereas YES Essentials was developed through an open-line brief based on internal knowledge and consumer research. It is not uncommon for the OEMs to tender briefs to more than one supplier at a time. OEMs distribute requirements via a brief to specify requirements in terms of objectives for the textiles; this process may commence up to 3 to 5 years in advance of the final vehicle production dependent on whether incremental or innovative design is involved. Dependent on the client's requirements, the brief may focus on pattern requirements and technological developments, whilst others may be more responsive to emotive desires of the consumer (Figure 3a,b).

It may be noted that an extra consideration needs to take place when developing a concept by open-line brief, in that the technical feasibilities and standards for the concept need to be researched; moreover, internal briefs are no longer common at Sage as the resource implications and risks to approval by a potential client are high and raise too many complex questions when considering how copyright is assigned and to whom (Sage or their client).

Stage A1: plan and conduct research and development

This initial stage A1 can take between 2 weeks and 2 months dependent on the client's needs. According to Powell, a brief will consist of performance [standards] and cost parameters, colour, pattern, scale of pattern, lustre and handle which is then taken forward by the designers (Powell 2008a). Timing and major stage-gates will also be confirmed. If the initial stage-gate is short, the design team will show concepts already *in situ* to support quick turnaround of sampling; when the stage-gate is long, the designers will develop new concepts for a particular brief, and this may include two stages prior to sampling.

Vehicle interiors are mainly plain, striped, checked or small-scale ditsies in neutral colouring with the addition of a small percentage of a brighter accent colour to add interest (Powell 2008a). Various OEMs who produce shorter runs have started to offer special editions to add more interest into the colour and design of the automotive (Powell et al. 2008b citing Powell 2006). As already stated, vehicles are developed up to 5 years in advance; alongside this is the timeframe of 7 years that the textiles will be in vehicle production before an updated model is developed and ultimately replaces it. Thus, the fabric has to remain on trend for a much lengthier amount of time than fashion or home interiors in terms of appearance and colour.

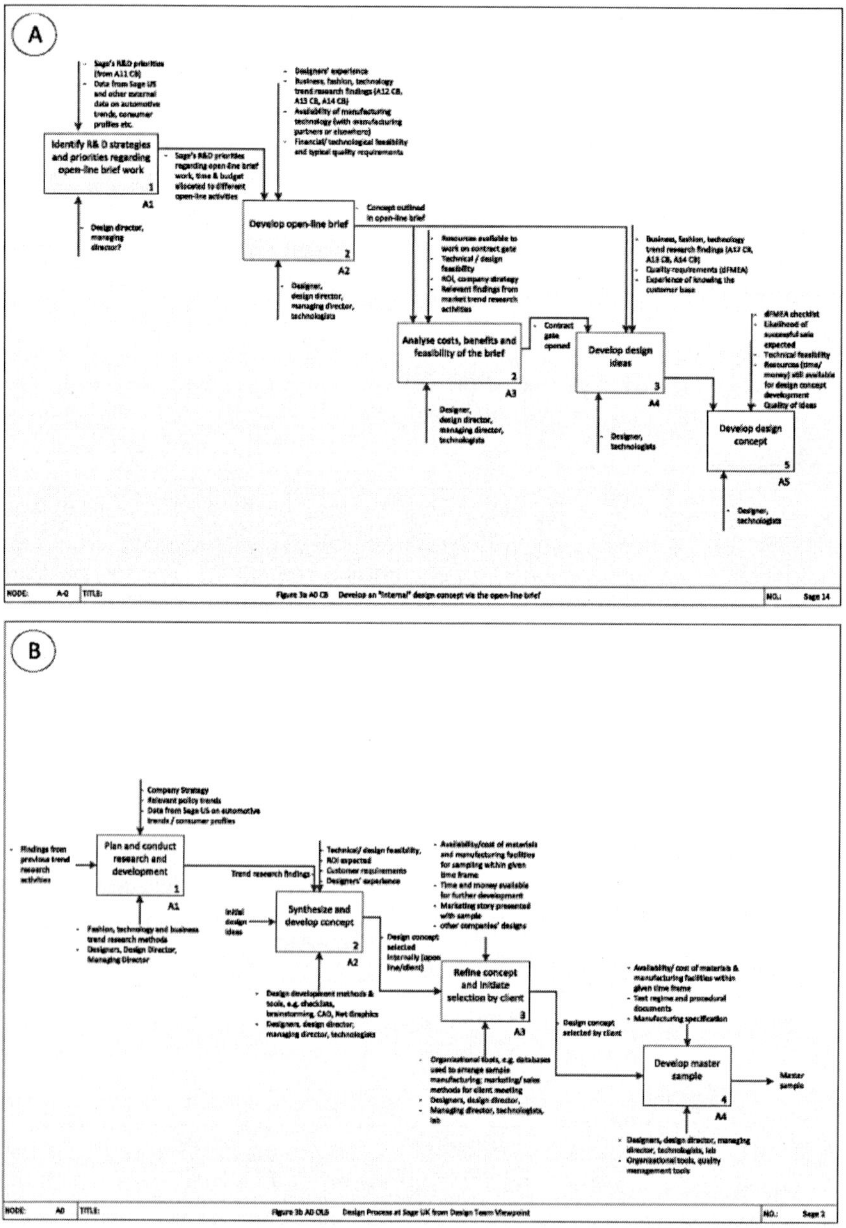

Figure 3. Open-line brief. (A) Open-line brief product development. **(B)** OEM set brief product development.

Due to the long lead times, a global macro-trends study is projected every 4 years and micro-trends analysed such as ‘luxury’, ‘sports’ and ‘comfort’ to focus the design team. Sage AI used to profile geographically but now references according to age and lifestyle, e.g. ‘youth’ and ‘active’. Other elements to be considered are environmental trends, colour, texture and consumer profile. The designers produce a global colour show consisting of four themes to maintain constant fabric design and development and visit trade shows to keep abreast of latest yarn or fabric development techniques and suppliers. Figure 4a illustrates the open-line brief route whilst 4b illustrates the OEM set route.

The development of an internal brief provides an opportunity for Sage to develop environmentally friendly innovations. Whether those are taken up by the car manufacturers depends on trends and politics, company strategies and product cost. The risks of the product being accepted are high at this stage; therefore, it is important for Sage to research trends, politics, potential clients’ company strategies and estimate the product’s cost and feasibility; all of this is resource intensive.

Stage A2: synthesis and development of concept

The initial idea is transferred into a CAD software. Figure 4c illustrates that the open-line brief route requires an examination of the return on investment of resources spent on developing the concept and its fit with the company’s strategic priorities before they proceed to developing it into design ideas for the client.

During stage A2, the design idea is developed into a concept (that will be discussed at a meeting with potential clients). At stage A24, the design idea is examined for selection with regard to whether to continue to refine or not. The design ideas might not be good enough (in which case, they need to go back one step). A design functional modes and effects analysis (dFMEA) review is undertaken which outlines the technical requirements (included in dFMEA) to narrow down the selection of potential fabrics and material; much of this depends on the design team’s knowledge of what is likely to work. Technological requirements are manufacturer dependent. Figure 4d illustrates the activities surrounding the decision to select the design concept to develop further.

At stage A25, the concept is refined towards developing a sample. The design concept (perhaps in the form of a CAD illustration, or a selection of archive samples), includes specification of materials, technologies required for the making, sources of laminators, etc.

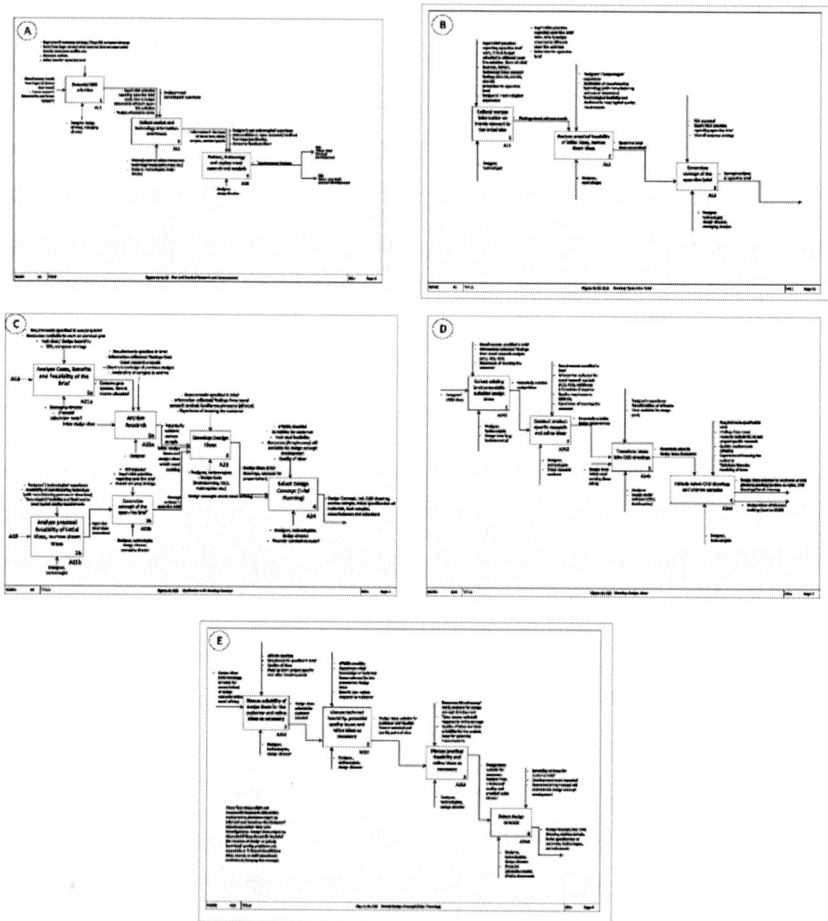

Figure 4. Stages A1 to A25. (A) Client based brief route for stage A1: plan and conduct research and development. **(B)** Open-line brief route for stage A1: plan and conduct research and development. **(C)** Stage A2: synthesis and development of concept. **(D)** Stage A24: developing the design idea. **(E)** Stage A25: select design concept (trial planning).

Figure 4e illustrates the activities involved in refining the idea into a concept that will be developed for discussion in a client meeting. The concept is produced on a CAD programme (in this case, NED Graphics); this is emailed to the specified manufacturing location and processors and (dependent on colourways) will take 4 to 6 weeks for sampling to materialize. If the colour of the yarn is already available (e.g. black), this would be a 4-week process; however, if the yarn needs to be dyed specifically to match other elements of the interior such as vinyls and leathers, this can take approximately 6 weeks.

Stage A3: refine concept and initiate selection by client

Concepts considered feasible for client discussions are developed into a sample representative of the actual fabric that may be used by the client company (Figure 5a).

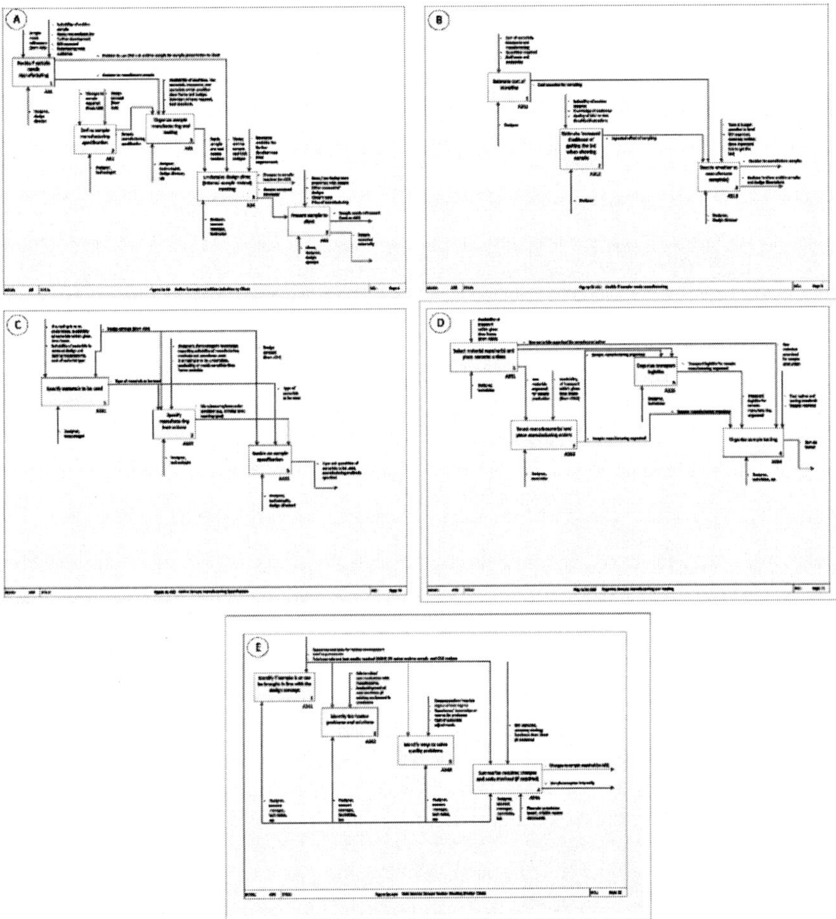

Figure 5. Stages A3 to A34. (A) Stage A3: refine concept and initiate selection by client. **(B)** Stage A31: initial decision if a sample should be manufactured. **(C)** Stage A32: defining sample manufacturing specification. **(D)** Stage A33: organizing sample manufacturing and testing. **(E)** A34: hold an internal review meeting (design clinic).

Samples to be considered may be new samples designed or they may be archival samples or samples that did not quite achieve the appropriate standards in a previous round of testing. The overriding considerations are return on investment (likelihood to get the job, likelihood to get good money out of job if accepted), availability of machinery and raw material within specified time frame, time availability and cost of sampling. The initial decision-making regarding whether a sample should be put into manufacture is illustrated in Figure 5b.

Once the decision is made to take a concept to the manufacturing level, specifications for manufacturing need to be developed and defined, as illustrated in Figure 5c.

As the sample specifications are being defined, there is also the need to organize the logistics for the transportation of the samples and their testing. Samples in automotive textiles are many metres of fabrics, representing bulk and hence the requirement for cost and time efficiencies to be met, as illustrated in Figure 5d.

Once tested, the sample is then examined by the design team for quality of manufacture, standard of testing and financial implications. Any problems are identified with appropriate rectification routes. Samples accepted through this internal design review are presented to the client company (stage A35) who will make the decision to either review the sample again after specified modifications or to agree that the sample is further developed for manufacture.

Stage A4: develop master sample

Once the clients have reviewed the samples, decisions are made regarding the merits of developing the sample for production; the first stage of this is the 'master sample'. Approved samples from the client review at stage A35 undergo a second set of testing and colour-tuning (Figure 6); if not approved, design alterations and further sampling will occur. When samples and testing are approved, the design process is finalized, a term described by Sage AI as 'mastering', and this is whereby the whole product is fixed (yarns, design, etc.), contracts are raised, the client firm is financially committed and the design is then assigned to production.

3.4 Decision making through the design process regarding recycled textiles

Sage AI has maintained sustainability as core to their corporate values from Milliken and normally use polyester yarns for their textile products that are visible within the automotive (because it is cost-effective, recyclable and meets the rigorous test requirements). The polyester may be coloured and texturized to a very wide variety of looks and feel that are cost-effective, visually appealing to the customer (the OEM) and pass a number of rigorous tests (such as abrasion and light fastness) so that it will last for up to 10 years.

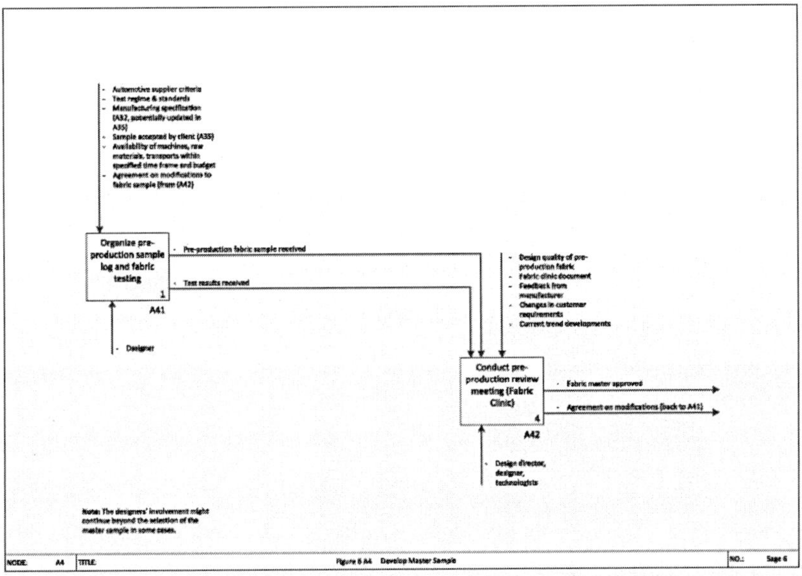

Figure 6. Stage A4: developing the master sample.

In line with their sustainability vision, Sage AI also use a percentage of post-industrial recycled yarns and post-consumer (PTC) recycled yarns blended with virgin yarns. Post-industrial recycled yarn is easier for Sage AI to use as properties and provenance are known. Most yarn manufacturing takes place in Asia, but the suppliers of polyester yarns are based in wider locations, e.g. Sage AI (UK) have a supplier in the UK in Nottingham. As the tests that automotive yarns and fabrics have to pass are so rigorous, there are few automotive yarn suppliers. Automotive textile interior fabrics and yarns are therefore sourced from a small pool of yarn suppliers, easing the complexity of the provenance of the yarns.

PTC recycled yarns are made from 100% plastic bottles, and the benefits of using post-consumer recycled yarns is the reduction in greenhouse gas emissions, the reduced energy consumption compared to making virgin yarns from raw materials and associated landfill charges and land use for storing waste plastic bottles (Collora 2012). The benefits are somewhat counterbalanced by the considerations of another supply chain loop; post-consumer yarns are currently 100% from plastic bottles which require processing into yarns, and there are the collection of plastic bottles, cleaning, transportation and storage of bottles before manufacturing processes, raising issues about transparency within this recycling route (European Environmental Agency 2012).

The use of recycled textiles used in production is a dilemma: the recycled polyester is blended with virgin polyester and so the resultant fabric does not look different from the non-recycled fabrics, thus raising questions from OEM customers and final vehicle consumers alike about the validity of the sustainable claims and extra costs. The cost of using recycled yarns appears to be in the processing of the yarn: there is a surcharge/premium in terms of the raw materials which is based on the polyester and not the plastics making it difficult for the OEM customer to understand the costing. The cost implications are, for example, 7% mark up on fabric that is 40% recycled and 11% on fabric that is 100% recycled; when this is taken back to the yarn, the cost is double or three times that on the normal price of the yarns. According to Ford's research shared with Sage AI, the premiums charged on recycled yarns varied across the world: 10% to 15% (USA), 15% to 20% (China, India, Brazil), 20% to 25% (Europe), 20% to 30% (Korea), and 50% to 100% (Thailand).

As well as the yarn composition, the application of textiles within the visible areas of automotives presents a problem regarding sustainability. Between seven to nine million vehicles are disposed of annually in the EU; only 75% of this is recyclable. Textiles form part of the remainder non-recyclable materials (Collora 2012). Textiles from ELV seating make up about 15% of automotive shredder residue, (ASR - possible to process into vehicular sound proofing insulation and then finally incinerated) (Erth et al. 2008a, 2008b).

Seat covering textile is laminated to impart waterproofness, increased abrasion, stain, flame and UV resistance, retro-reflection properties which the OEM requires testing standards to be achieved. Lamination involves backing onto foam, usually with adhesive. Currently, the most cost-effective lamination process renders textile products to be non-recyclable as lamination involves placing adhesives onto the fabric. It is possible to replace adhesive and foam-based lamination process with three-dimensional knitted or non-woven textiles products stitch-bonded or needle-punched onto the exterior fabric enabling separation of layers and recyclability. It is expensive though and currently only in high-value cars.

As costs appear to be an overriding consideration, we examined the design process to identify where and how decisions were made that were directly related to considerations about using recycled yarns (and by implication the lamination process used). This is presented in Table 1.

From Table 1, it can be seen that there are few, if any, decisions that are under Sage AI's control, save the decision to develop an open-line brief. This is a route that is not now very common for Sage AI as it is risk-laden and demands resources in terms of time and money with no guarantees of developing a sample that wins a contract. However, having developed fabric samples in this way to show at various industry shows did place them as companies who are not only interested in sustainability but also actively seeking solutions.

Decisions around developing sustainable textiles solutions for automotives are through negotiation with the lead firm, the OEM that is seeking to manufacture the model. A study of incorporating sustainability into the supply chain management at Volkswagen (Koplin et al. 2007) noted essentially two forms of environmental supply management: (1) the integration of environmental criteria/standards into product and production-related decisions along the whole supply process ('greening the supply chain') and (2) the optimization of the environmental compatibility of purchased goods ('product-based green supply') (Vermeulen et al. 2011).

In the first route, the supplier is evaluated for their environmental performance; achieving and providing certificates for environmental management systems such as ISO 14001 are regarded as minimum qualification criteria for discussions to commence about supplying a product. Sage AI (UK) has ISO 14001 and will also work with suppliers of their raw materials who have either ISO 14001 or QS-9000 (an automotive standard developed by General Motors, Chrysler and Ford based on ISO 9000). In the second route, all stages of the life cycle are considered from packaging, recycling and disposal. As products are developed in accordance to specifications set by the OEM in the brief, measurable sustainable criteria should help to evaluate the environmental credentials of the product. However, as the case study with Sage AI (UK) attests, the interest in green procurement of sustainable textile products for the automotive sector in Europe has not gained enough ground to warrant all or majority of textile design to be from recycled polyester or the use of non-adhesive lamination. A recent study of sustainable supply chain management within the Indian textile industry not only noted that it was important to identify the most influential drivers within the region as these enablers were observed to be context based (Diabat et al. 2014) but they also observed that within the automotive sector, the customer was a very influential driver to sustainable textile design.

Table 1. Decision-making for using recycled yarn

Decision	Decision explained	Relevance for environmental sustainability	Relevance in recycled yarn scenario?
Determination of R&D priorities (A11)	R&D priorities (including the level of detail at which this is defined), time and budget allocated to different open-line activities	Level of priority given to EPD in R&D has direct implications for both innovative and incremental EPD activities (incl. level of detail about specific EPD aspects)	N/A
Decision whether to work on a company brief or not (contract gate opening) (A21a)	Decision whether to open a contract gate is made; a certain amount of time and money is allocated to the project	a) The decision whether to work on a brief which requires research into EPD is made, and the time and money allocated to the brief is determined (impacts on the likelihood of the bid being successful). b) The amount of time/ money allocated to a 'conventional bid' affects the time left for work on specific EPD projects. c) For any bid, the availability of resources (amount of time/ number of staff members allocated to working on the bid) could influence whether there is scope for exploring EPD as part of the design development	Yes
Decision on which open-line briefs to take forward (internal product development work) (A21b and A21b)	The decision which open-line brief ideas to work on is taken/ open-line brief ideas are concretized	Taking the R&D priorities into account, the process of narrowing down design ideas requires a decision on whether to work on EPD as part of the R&D work - and what priority they are given over other R&D ideas - this may be an implicit or explicit process	N/A
Design idea development - decision which ideas to present (A23)	Design ideas are selected (CAD drawings, samples for presentation)	A decision whether to include design ideas which require sustainable materials and production methods is made.	Yes (combined analysis with design concept selection)

Design concept selection (A24)	Design concept selected, incl. CAD drawing, archive sample, initial specification of materials, technologies, manufacturers (A4 selection)	The design concept selection will largely determine what the actual product will look like and how it will be made, which has direct implications for the environmental impact of its production.	Yes (combined analysis with design idea development)
Organization of sample manufacturing and testing (A33)	a) Suppliers and manufacturing partners are selected for the product sample (if successful, they are likely to be commissioned with the production of the product once it goes into production as well). b) Testing is organized (test regime specified by the respective OEM)	The choice of suppliers and manufacturing partners will affect the environmental impact of the product. The importance of the choice of the manufacturing partners depends on the availability of alternatives and the variability between the manufacturing processes with regards to their environmental impact (e.g. the age of the machines used, materials used, energy sources etc.) - the differences are likely to be greater if the market is global.	Yes

4. CONCLUSION

This paper presents the dichotomies in developing sustainable textile materials for the automotive industry. Recycled and renewable raw materials are two main drivers in the growth of the Mobiletech markets which accounts for about 22% of the technical textiles base, making it the largest single market and a growth area. There is, however, a disparity in the use of recycled yarns in automotives: in the US market, the Ford Mondeo Fusion Hybrid has been reported as having 100% recycled yarn whilst the model in the Asian and EU markets is reported to have up to 43% recycled yarn (Green Supply Chain News 2013). According to Messe Frankfurt (2013), the use of textiles within automotive has risen from 20 kg per car in 2,000 to 26 kg in 2012 and is set to rise to 35 kg in 2020, driven partly by the use of textiles composites and substitution of heavier materials for textiles to cover fuel consumption and CO2 emissions. Sustainability is important for consumers in the mature automotive markets of the USA and EU; of the top ten decisions driving purchase of automobiles, fuel efficiency was number one and that environmental considerations was number four (KPMG 2013).

As the use of recycled fabrics (and the use of any recyclable textile component) push up the final costs of the car, the consumer has to be interested to buy recycled/sustainable, and this is not easy to predict: there is a large interest in the North Americas, but not so much of a concern within Europe. Research conducted by (European Environmental Agency 2012) identified that in their sample of 1,000 men and women in USA, 70% of them felt that sustainability was now a mainstream issue for them and that 9 out of 10 of these ‘green consumers’ wanted to know where recycled materials are used and what recycled fibre brands are used in their cars. The decision to use recycled content within the product therefore rests on cost and questions that need to be addressed by Sage AI (UK) which include:

- Is it a ‘deal-breaker’, i.e. would the use of recycled textiles parts secure a sale?
- What are the cost implications on the total supply chain?
- Does legislation present any advantages in the use of recycled textiles in production?

Our study demonstrates how the design for automotive textiles is a system, whereby design decisions are a series of negotiations with the client OEM. Our interviews found that although Sage AI headquarters did provide worldwide trends information to all locations, there was some concern regarding consumer knowledge about the EU markets; the implications for Sage AI (UK) are most notable in their ability to persuade the client to work with them towards the most sustainable textiles solutions, rather than achieving lowest cost. With reference to Meadows (1997) and her nine interventions for influencing and changing systems, the automotive sector already have governmental policy drivers and material stocks and flows in that the technology is currently available for recyclable and sustainable textiles to be developed for the automotive sector. Given the strength of consumer interest in the US for sustainable textiles within the vehicle, it would be interesting to research how the US consumer became engaged with this. Some of the questions raised from this case study are as follows:

- How are the costs of recycled yarns determined?
- How can equality in costs of recycled/virgin yarns be achieved?
- How can/should the sustainability message for recycled textiles be delivered?
- What new technologies may be developed to ensure that end-of-life considerations can be taken into account when developing laminated textiles for automotive interiors?

COMPETING INTERESTS

The authors declare that they have no competing interests.

REFERENCES

1. Charnley, F., Lemon, M., Evans, S. (2011), Exploring the process of whole system design, Design Studies 32 (2011) 156–179,

2. Collora, J. (2012) Recycling empty PET bottles, Bottled Water Reporter, December 2011-January 2012, pp8-14, International Bottled Water Association http://www.nxtbook.com/ygsreprints/ IBWA/ g23199_IBWA_DecJan12/#/10

3. Diabat, A, Kannan, D, & Mathiyazhagan, K. (2014). Analysis of enablers for implementation of sustainable supply chain management - a textile case. *Journal of Cleaner Production, 83*(2014), 391–403.

4. EEA (European Environmental Agency) (2012) Movements of waste across the EU's internal and external borders, EEA, ISBN 978-92-9213-337-5

5. Erth, H, Gulich, B, & STFI. (2008a). Three-dimensional textiles and nonwovens for polyurethane foam substitution in car seats. In R Shishoo (Ed.), *Textile advances in the automotive industry* (pp. 140–149). Cambridge: Woodhead Publishers Ltd.

6. Erth, H, Gulich, B, & STFI. (2008b). Three-dimensional textiles and nonwovens for polyurethane foam substitution in car seats. In R Shishoo (Ed.), *Textile advances in the automotive industry* (p. 147). Cambridge: Woodhead Publishers Ltd.

7. EURAPA. (2010). Summaries of EU legislation: end-of-life vehicles. Retrieved 07-11-2011, from http://europa.eu/legislation_summaries/environment/waste_management/l21225_en.htm

8. Green Supply Chain News (2013) Ford Motor Company keeps pushing the green envelop as it continues to expand use of recycled materials inside its cars, http://www.thegreensupplychain.com/news/13-05-20-1.php?cid=7093

9. Hernandez-Matias, JC, Vizan, A, Perez-Garcia, J, & Rios, J. (2008). An integrated modelling framework to support manufacturing system diagnosis for continuous improvement. *Robotics and Computer-Integrated Manufacturing, 24*(2), 187–199.

10. Hirao, M., Sugiyama, H., Fischer, U., & Hungerbühler, K. (2008). IDEFO activity modeling for integrated process design considering environmental, health and safety (EHS) aspects. In B. Bertrand & J. Xavier (Eds.), Computer Aided Chemical Engineering (Vol. Volume 25, pp. 1065–1070): Elsevier.

11. Kikuchi, Y., & Hirao, M. (2009). Activity and information modeling of comprehensive assessment for sustainable process design. In J. Jacek & T. Jan (Eds.), Computer Aided Chemical Engineering (Vol. Volume 26, pp. 1123–1128): Amsterdam, London:Elsevier.

12. Kikuchi, Y., Papadokonstantakis, S., Banimostafa, A., Sugiyama, H., Hungerbuhler, K., & Hirao, M. (2012). Analysis and modeling of information required for process assessment on environment, health, and safety by IDEF0 and UML. In A. K. Iftekhar & S. Rajagopalan (Eds.), *Computer Aided Chemical Engineering* (Vol. Volume 31, pp. 1392–1396): Amsterdam, London:Elsevier.

13. Knowledge Based Systems, I., (2010). IFED0 - function modelling method, Retrieved 17/09/2012, 2012, fromhttp://www. idef.com/ IDEF0.htm

14. Koplin, J, Seuring, S, & Mesterharm, M. (2007). Incorporating sustainability into supply management in the automotive industry - the case of the Volkswagen AG. *Journal of Cleaner Production, 15*, 1053–1062.

15. KPMG. (2013). *KPMG's global automotive executive survey 2013.* https://www.kpmg.com/KZ/ru/IssuesAndInsights/ArticlesAndPublications/Documents/KPMGs-Global-Automotive-Executive-Survey-2013.pdf

16. Meadows, D., H, (1997) Places to intervene in a system, Whole Earth, Winter, http://www.wholeearth.com/issue/2091/

17. Messe Frankfurt (2013) Sustainability Recycling concepts and natural fibres in motor vehicles,http://www.messefrankfurt.com/ frankfurt/ en/ media/textiletechnologies/techtextil/frankfurt/texte/autositze-und-dachhimmel-aus-holz--flachs--mais-und-recycel-tem-.html

18. Mingers, J, & White, L. (2010). A review of the recent contribution of systems thinking to operational research and management science. *European Journal of Operational Research, 207*, 1161. Elsevier.

19. National Institute of Standards and Technology. (1993). *Draft federal information processing standards publication 183 - standard for integration definition for function modeling (IDEF0).*

20. Normand, X. (2008). In R Shishoo (Ed.), *Textile advances in the automotive industry* (pp. 86–109). Cambridge: Woodhead Publishing Ltd. pp. 87–88.

21. Powell, NB. (2006). New textile product development: processes, practices, and products. *The Journal of The Textile Institute, 97*(2), 155–166.

22. Powell, NB. (2008a). Design of automotive interior textiles. In R Shishoo (Ed.), *Textile advances in the automotive industry*(pp. 113–139). Cambridge: Woodhead Publsihing Ltd.

23. Powell, NB, Handfield, R, & Barnhardt, R. (2008b). Mapping the automotive textile supply chain. In R Shishoo (Ed.), *Textile advances in the automotive industry* (pp. 17–42). Cambridge: Woodhead Publishing Ltd.

24. Rashid, MM, & Ismail, H. (2007). Generic tool for measuring the reliability of product development processes. *Journal of Modelling in Management, 2*(1), 71–93.

25. Shishoo, R. (2008). *Textile advances in the automotive industry*: Woodhead Publishing.

26. Sugiyama, H, Hirao, M, Fischer, U, Hungerb, U, & Hler, K. (2008). Activity modeling for integrating environmental, health and safety (EHS) consideration as a new element in industrial chemical process design. *Journal of Chemical Engineering of Japan, 41*(9), 884–897.

27. Taylor, I. (2012). *Interview with design and product development director at Sage AI, December 2011.*

28. Vermeulen, I, Van Caneghem, J, Block, C, Baeyens, J, & Vandecasteele, C. (2011). Automotive shredder residue (ASR): reviewing its production from end-of-life vehicles (ELVs) and its recycling, energy or chemicals' valorisation. *Journal of Hazardous Materials, 190*(1–3), 8–27.

29. Yin (2009), Case study research: design and methods, p. 13, California:Sage

30. Yu, B, & Wright, DT. (1997). Software tools supporting business process analysis and modelling. *Business Process Management Journal, 3*(2), 133–150.

CHAPTER 7

Wearable Wireless Cardiovascular Monitoring Using Textile-Based Nanosensor and Nanomaterial Systems

Prashanth Shyamkumar [1,*], Pratyush Rai [1], Sechang Oh [1], Mouli Ramasamy [1], Robert E. Harbaugh [2] and Vijay Varadan[1,2,3]

[1] Department of Electrical Engineering, University of Arkansas, Fayetteville, AR 72701, USA

[2] Department of Neurosurgery, College of Medicine, Pennsylvania State University, Hershey Medical Center, 500 University Drive, Hershey, PA 17033, USA

[3] Global Institute of Nanotechnology in Engineering and Medicine Inc, 700 Research Center Blvd, Fayetteville, AR 72701, USA

ABSTRACT

Wearable and ultraportable electronics coupled with pervasive computing are poised to revolutionize healthcare services delivery. The potential cost savings in both treatment, as well as preventive care are the focus of several research efforts across the globe. In this review, we describe the motivations behind wearable solutions to real-time cardiovascular monitoring from a perspective of current healthcare services, as well as from a systems design perspective. We identify areas where emerging research is underway, namely: nanotechnology in textile-based wearable monitors and healthcare solutions targeted towards smart devices, like smartphones and tablets.

KEYWORDS

nanotextiles; smart textiles; electrocardiography; wearable healthcare; cardiovascular disorders; ambulatory monitoring; dry electrodes; conductive nanocomposites; smartphone

1. INTRODUCTION

Advancements in pervasive computation and communication technologies, coupled with micro- and nano-electronics have created opportunities for the seamless integration of electronics and flexible sensors. These flexible sensors made of textiles or possessing a clothing-like texture and flexibility result in functionalized textiles, commonly referred to as e-textiles or smart textiles. Tao [1] describes smart textiles as a class of smart materials and structures that sense and react to environmental conditions or stimuli. Depending on the degree to which intelligence is imparted into these textiles, they may be passive, active or very smart. Passive textiles only acquires information about the environment [2]; active textile reacts to its environment and adapts in different ways [3]; and very smart textiles may be context-aware and adapt their responses based on the context [4]. These three components may be confined to the presence of sensors, actuators and controlling units. For instance, a passive material requires sensors to detect signals. Constituents of an active material may require the coupling of an actuator or a controlling unit with a sensor. Similarly, for very smart textiles, a combination of sensors, actuators, controlling units and a context-aware material that functions as a processing unit may be essential. Context-aware intelligent materials are formulated by combining conventional textiles with novel materials, mechanics, processing techniques and the chemistry and biology of materials [5].

In biomedical applications for long-term physiological signal monitoring, textiles are the preferred platform for sensors, because they are the most natural materials close to the skin. Traditionally, long-term ambulatory systems that perform biomedical monitoring use adhesives to hold the electrodes or sensors in place against the skin and use a conductive gel to make low impedance electrical contact with the skin. Advancements in analog front-end electronics with very high input impedance and a high common mode rejection ratio (CMRR) have enabled the use of dry electrodes without the need for gels [6]. These dry electrodes when made on textiles can be seamlessly integrated into clothing to achieve an intuitive and comfortable alternative to currently available long-term monitoring systems. However, these dry electrodes have been known to have an initial settling time, wherein the contact impedance slowly drops to a value adequate to acquire signals with minimal noise from the environment [7]. One approach to alleviate this problem is to use nanostructured sensors, which have significantly higher surface area, as compared to planar sensors, so that the desirable low contact impedance can be achieved sooner. In this paper, our focus is on the use of nanomaterials and nanostructures in textiles for wearable biomedical monitoring systems for cardiovascular health.

The review is organized as follows: First, we present the motivations for a textile integrated wearable system from a healthcare perspective for patients who have already been diagnosed with a cardiovascular disease (CVD) and for occult conditions that pose sudden and high risk to individuals under physical exertion. Second, we describe the research performed thus far on the use of nanostructured sensors and nanomaterials on textiles for cardiac monitoring. Finally, we present examples of end-to-end system implementations for multi-parameter monitoring through smart devices, namely the e-bra for women and the e-bro for men.

1.1. Cardiovascular Disease Patients

In the U.S., the cost of healthcare has increased by over 90%, from $1337 billion in 2000 to $2594 billion in 2010. It is projected to reach $4,487 billion in 2020 [8]. Public spending on health and long-term care in Organization for Economic Co-operation and Development (OECD) member countries and BRIICS (Brazil, Russia, India, Indonesia, China, South Africa) is 6% of the gross domestic product (GDP) and is projected to increase up to 14% in the next 50 years [9]. Chronic disease diagnosis and treatment are the primary causes for this increase. Patients suffering from chronic diseases need to repeatedly visit the hospital, which can be expensive. As a solution to this, remote point-of-care (POC) systems and remote patient monitoring (RPM) systems can be used. Remote patient monitoring for point-of-care facilitates the monitoring of a patient's health condition at local or remote places without the need for hospital admission or visits. In the case of high risk patients, it can provide the patient with real-time feedback from a medical center.

1.2. Athletes: Football and Soccer

The sudden and fatal failure of an individual's heart function is referred to as sudden cardiac death (SCD). It is well known that the risk of SCDs and acute myocardial infarctions (AMI) increase during and immediately after strenuous exercise [10]. Nearly 58% of SCDs reported between 1980 and 2006 have been reported in basketball and football athletes. Recent studies have shown that the incidence of SCDs in young athletes in the U.S. at the high school and college level have been underestimated in previous studies [11]. The most prevalent causes of SCD in young athletes are CVDs and sports-related injuries. Among CVD causes, the most prevalent are hypertrophic cardiomyopathy (HCM) (36% of cases) and coronary artery diseases (CAD) (17% of cases). Among sport injuries, commotio cordis and blunt trauma injuries together account for 25% of all SCDs recorded between 1985 and 2006 [11]. The current strategy for the prevention of SCDs in young athletes is to prescreen them and diagnose any cardiovascular diseases that may put them at high risk for SCDs and promptly disqualify them from participation if diagnosed. The proven approach implemented in Italy has involved a mandatory prescreening with detailed history, physical examination and a 12-lead ECG with guidelines and criteria for the identification of cardiovascular abnormalities that may put the athlete at high risk for SCD [12]. The American Heart Association (AHA), however, does not currently recommend the inclusion of 12-lead ECG as a part of the prescreening [13,14,15] for several reasons: the high direct costs of the tests, a lack of dedicated trained athletics personnel to perform the prescreening in place of physicians, the sheer number of athletes to be screened and the reported low specificity, high false positives and false negatives of ECG interpretations [16]. Although the positive diagnostic value of including a 12-lead ECG in prescreening has been identified by both the European Society of Cardiology (ESC) and the AHA [17,18], consensus panels for recommendations on cardiovascular screening of young athletes, the cost-effectiveness of including this test in the U.S. athletic prescreening protocol is still a subject of wide debate.

Despite the evidence suggesting the effectiveness and initial success of prescreening with ECG in Italy [12], there are four limitations to the prescreening approach that need to be addressed:

First, there is still a wide debate on the differential diagnosis of HCM from the ECG changes brought on by training in many athletes with otherwise normal hearts (athlete's heart) [19,20]. Reported differences in training-induced cardiac remodeling between athletes of African origin and others have made diagnoses based on ECG findings equivocal [21]. Moreover, the remaining two prevalent causes for SCD (CADs and blunt trauma injuries) cannot be diagnosed during prescreening, as CADs do not manifest as ECG abnormalities and blunt trauma injuries are non-pathological and can occur in any athlete with an otherwise healthy heart.

Second, recommendations [22] suggest that for the differentiation of HCM from athlete's heart, Brugada-like ECG abnormalities, arrhythmogenic right ventricular cardiomyopathy or dysplasia and features, like prolonged PR intervals, short PR intervals, early repolarization and inverted or biphasic T waves, can be further evaluated using an exercise test to improve the specificity. However, this is to be done in addition to the preliminary ECG screening at an added cost.

Third, from the perspective of secondary prevention, *i.e.*, through the adoption of strict guidelines on sudden cardiac arrest (SCA) resuscitation, it is imperative that an SCA is promptly recognized, cardiopulmonary resuscitation (CPR) is started immediately and a defibrillating shock is applied as soon as possible [23]. The target resuscitation time recommended by the AHA is between 3 and 5 min, from the time the athlete's collapse was witnessed to the application of the defibrillating shock. It has been shown that survival chances may drop by 7%–10% for every minute that defibrillation is delayed. In the absence of a real-time ECG, the emergency responder or rescuer has to first identify an SCA with accurate pulse or respiration assessments, while the athlete may be gasping or having myoclonic jerks or seizure-like activity, which may be inconsistent with an SCA.

Fourth, the various mechanisms for SCD have been studied extensively at the cellular process and ionic channels level [24]. This work needs to be augmented with real-time studies on the mechanism of SCD using non-invasive techniques, like ECG, which are lacking. The ECG is rarely or never available during a sudden cardiac arrest episode. Therefore, a system for the real-time monitoring of cardiac electrophysiology during exertion, which puts the athletes at higher risk of SCDs, is an important step in the prevention and treatment of sudden cardiac arrest in athletes.

1.3. Military Recruits under Training

Among U.S. military recruits between the age of 18 and 25 years, the incidence rate of non-traumatic SCDs were found to be 11.1:100,000 in a 25-year period [25]. Recently, a cohort study of active military personnel who died on duty showed that 92% of the sudden death occurrences during military training were during organized physical training. According to the same study, 31.6% of the sudden deaths occurred during running, which was part of the organized physical training. The study further concluded that among persons ≥35 years of age, more emphasis should be placed on evaluating the risk of atherosclerotic coronary disease [26].

Based on the three scenarios described above, we can formulate a system requirement as follows:

- An ergonomic, intuitive and wearable, fully-integrated platform to acquire ECG and blood pressure in real time.
- Wireless communication to transfer the acquired biomedical signal in real time to a data logging device or a computational device that does real-time analysis.

2. SENSORS ON TEXTILES

Smart textile (fabric) can be made from materials ranging from traditional cotton, polyester and nylon, to advanced Kevlar with integrated functionalities. However, in the scope of the present review, fabrics with electrical conductivity are of interest. There are two kinds of smart textile (fabric) products that have been developed and studied for health monitoring fabric with textile-based sensor electronics [27,28,29,30,31] and fabric that envelopes traditional sensor electronics [32,33]. Pioneering research work, done by Jayaraman and co-workers [27,28], showed that weaving can be used to incorporate electrically-conductive yarn into fabric to obtain a textile that can be used as a "Wearable Motherboard". It can connect multiple sensors on the body, such as wet gel ECG electrodes, to the signal acquisition electronics. Later research has shown that conductive yarns can be instrumental in the fabrication of textile-based sensors made of fabric [29,30] or metallic meshes [31] coated with silver or conductive metal cores woven into the fabric [34].

There are two broad approaches to the fabrication of garments with ECG sensor electrodes in research:

- Finished garments through functionalization or integration of finished garments with sensor elements. This approach involves the integration of finished electrodes into finished garments by simply stitching the electrodes at the appropriate locations on the garment or using deposition techniques to transfer the functional materials at the appropriate locations.
- Unfinished garments—the introduction of smart materials during the garment fabrication process. This approach entails the use of textile fabrication techniques to form woven or non-woven fabrics with the inclusion of functional materials.

2.1. Incorporation in Finished Garments

Several flexible and rigid materials have been fabricated and evaluated for use as electrodes. Among resistive electrodes are flexible polydimethylsiloxane (PDMS) [35], CNT array electrodes named ENOBIO [36], carbon nanotube (CNT)/PDMS nanocomposites [37], flexible polymeric dry electrodes [38] and vertically-aligned metallic nanowires on flexible substrates [39].

Among capacitive electrodes are Ti/TiN electrodes [40], IrO-coated electrodes [41] and MEMs spiked electrodes [42]. A comprehensive review of these contact and non-contact dry electrodes has been presented in [43]. Smart textile implementations can be achieved by stitching these electrodes onto finished garments.

2.2. Textiles as Electrodes: Unfinished Garments

The smallest units of the textile are fibers or filaments. Innumerable combinations of these units can result in many textile materials with varying length, cross-section areas and shapes and surface roughness. The intelligent functionality, conductivity in this case, can be introduced into textiles at different levels. At the fiber level, a coating can be applied or conductive threads can be added to make a composite textile. Fibers of different types can be arranged at random or in a strictly organized way in yarns or fabrics to form even 3D structures. These structures can be metallized or functionalized to fabricate a conductive textile electrode and other functional surfaces with micro- or nano-rods, or micro- or nano-coil arrays. Accordingly, there are two strategies followed to make fabrics conductive: inclusion of thin conductive filaments in the yarns used to make the fabric or coating of the finished fabric with conductive materials by various coating techniques.

The former approach has been studied using stainless steel [44]. Mechanical, as well as conductive property enhancement has been documented with the absorption of single-walled carbon nanotubes (SWCNT) in cotton [45].

The latter approach uses deposition and coating techniques, like sputtering, screen printing, electro-spinning, carbonizing and evaporation deposition. Conductive coatings on fabric result in higher conductivities, but are less durable, especially through wash cycles [34].

Nanotextiles are formed by the fabrication of nanostructures on textile fabrics or forming nanoscale filamentous structures on fabric, rather than incorporating nanomaterials into fabrics. InOh *et al.* [46] have fabricated electrospun silver-plated polyvinylidene fluoride (PVDF) nanofibers for use as long-term dry electrodes. Vertically aligned nanostructures using a fabric flocking technique have been fabricated and tested by Rai *et al.* [47].

2.3. Nanocomposite Inks for Conductive Traces and Connections

Implementations of wearable systems on textile require a cost-effective, as well as scalable way to form interconnects between the various components. One approach to achieving this goal is to have functional inks that can be printed onto a textile to essentially form printed circuit boards on textile. In the case of ECG smart textiles, this involves the formulation of conductive inks or printable capacitive structures on fabric. Polymer thick film (PTF) technologies have been used to form conductive traces and transmission lines on a non-woven fabric in [48]. Alternatively, inks with conductive fillers and polymer binders have been used in the formulation of several inks that are stretchable and resistant to wear and tear. Stretchable conductive inks on textiles have been demonstrated through conductive traces using silver flakes in polyurethane-based binder by Araki *et al.* [49]. Poly(3,4-ethylenedioxythiophene):poly(styrenesulfonate) (PEDOT:PSS) films have been successfully deposited on rubber latex substrates using ink-jet printing by Romaguera *et al.* [50].

Multi-wall carbon nanotubes (MWCNTs) and polyaniline nanoparticle (PANP) core shell-based nanocomposite conductive inks were synthesized and successfully screen printed on woven cotton by Rai *et al.* [51]. Conductive traces drawn from inks with silver flakes in an acrylic binder have been used to make connections between ECG electrodes and flexible printed circuit boards (PCBs) [52].

In addition to conductive materials for resistive-type ECG electrodes, a flexible capacitive-type ECG electrode structure consisting of conductive CNT-acrylic nanocomposite inks sandwiched between two layers of acrylic inks has been fabricated and tested for ECG acquisition by Kumar *et al.* [53].

3. EXAMPLES OF SYSTEMS IMPLEMENTATION

Wearable systems for monitoring physiological signals have been reported in previous studies. Some of them are in the form of wearable accessories, such as wrist watch for cuffless blood pressure monitoring [54]. Systems can be off-the-shelf electronics with dry electrodes fabricated on FR4 boards [55] that can be embedded/encapsulated in clothing, such as a cap or shirt. The systems either display only or transmit to a remote location through Bluetooth-enabled cellular phone/Zigbee-enabled PC and wireless data transmission or the Internet for remote patient monitoring (RPM) [56,57]. The systems can be microelectronics fabricated on flexible substrates, which are easy to incorporate into the textile [58,59,60]. These electronics can accomplish multiple biometric monitoring, such as sweat rate, temperature, ECG, blood oxygen level, *etc.*, along with wireless transmission. In combination with global positioning technology, they can facilitate medical intervention and augment ubiquitous healthcare [59].

Sensors that are encapsulated, enmeshed or textile-based are predominantly electrical, optical or piezoelectric monitor physiological signals close to the body. Dry electrodes are the most popular type of sensors for biopotential measurement in wearable systems. These electrodes have varying designs, which include fabric made of woven conductive thread, an array of metalized needles, a conductive surface and an Ag/AgCl back electrode (conductive gel free) [61] and capacitively-coupled dry electrodes [62]. They have been shown to acquire biopotential signals. In addition to that, they have been proposed for cardiac, as well as neural stimulation. The textile-based dry electrodes have also been demonstrated as bioimpedance spectroscopy devices for plethysmography monitoring [63]. Optical sensors are mainly off the shelf microelectronics or optical fiber-based sensors. The piezoelectric sensors are off-the-shelf microelectronics, flexible piezoelectric membranes or functionalized piezoelectric yarn.

Wireless sensor systems generate large amounts of data that require data management and post processing. Research in database architecture has shown that database organization for the derivation of medical parameters for diagnostics and archiving of diagnostic parameters solves the problem of data volume [64]. In addition to that, data processing through cloud computing and remote access to stored data improves the computing performance [65].

In the following sections, implementations of wearable systems with textile-based nanosensors and state-of-the-art wireless communication systems are described.

3.1. E-Bro

The textile platform in this application is an inner vest that can incorporate nano-biosensors, such as gold nanowire electrodes [66] or nanostructured textile electrodes [67] and composite piezoelectric films [68,69]. It can also incorporate an infrared emitter-detection system for plethysmography and temperature sensors. The e-bro system is an implementation of a multichannel wearable wireless textile-based nano-biosensor that monitors ECG and blood pressure.

3.1.1. System Description

Figure 1 shows the overall system, which consists of four components: firstly, a compression inner vest, referred to as the e-bro, with the textile electrode sensors and the printed connection traces that connect the electrodes to a sensor electronics module (SEM); secondly, a photoplethysmography arm band that has near-IR LEDs and photodiodes, which are connected to the SEM through conductive traces printed from the left arm; thirdly, the SEM that consists of an amplifier, filter circuits, a microcontroller and a ZigBee wireless radio; lastly, a software program running on a PC that receives and plots incoming data from the person wearing it.

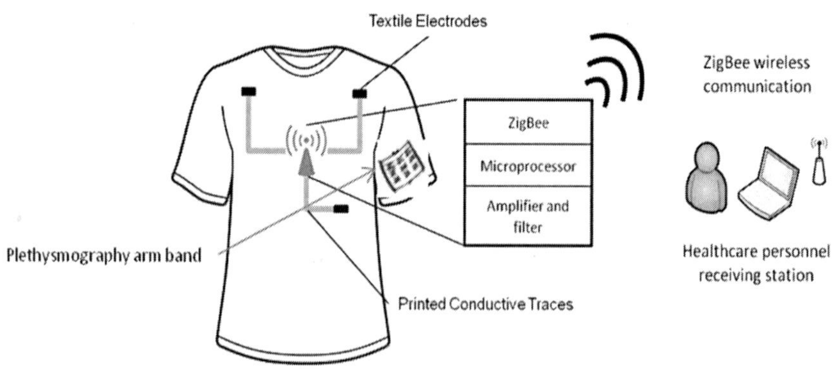

Figure 1. Schematic showing overall system for the acquisition of ECG through the e-bro.

The photoplethysmography arm band consists of two arrays of near-IR photodiodes and a central array of three photodiodes. This assembly was described in detail in [52]. The electrodes and the arm band are connected to the SEM through conductive traces, an ink formulation of silver nanoparticle fillers in an elastic acrylic-based binder, printed on the vest, and snap buttons. The conductive traces were made using conductive inks, and screen printing was used, which is textile manufacturing compatible. The arm band is removable and simple snaps on to the vest through four snap buttons (Figure 2). The removable SEM and arm band makes the e-bro washable.

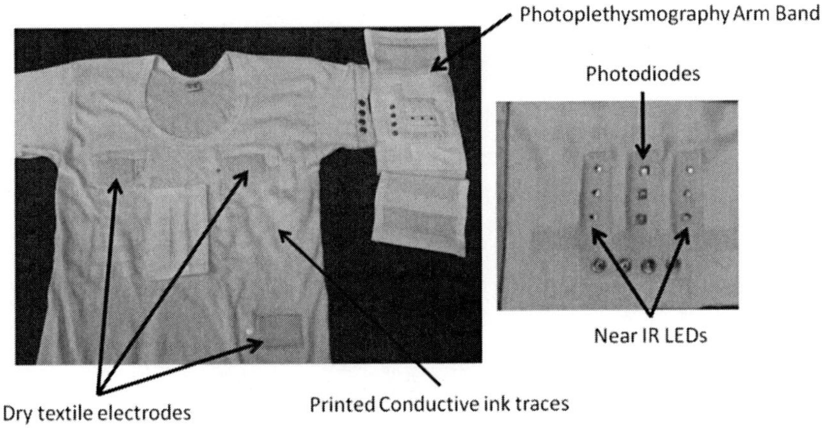

Figure 2. E-bro and the photoplethysmography arm band. Reproduced with permission from [47]. Copyright 2014, The Electrochemical Society.

3.1.2. Sensor Electronics Module (SEM)

The amplifier, which is a part of the SEM, consists of four channels: three channels for the bipolar limb leads, Lead I, Lead II and Lead III, and the fourth channel amplifies the potential difference across the photodiode, which detects the reflected IR waves from the brachial artery. The amplifiers used in the SEM had a band pass of 0.2 Hz to 70 Hz and a mid-band gain of 50 dB for the three ECG channels. The gain was increased to 55 dB for the photoplethysmography sensors for a band of 0.2 Hz to 15 Hz. The amplified signals from the amplifier are digitized using the onboard microcontroller for transmission; Figure 3.

The choice of the ZigBee radio module was motivated by two desired functions. Firstly, it has to support data rates higher than 9,600 bps, because four channels of digitized ECG and BP signal have to be transmitted in real time. Secondly, it provides communication ranges as high as possible for applications in sports, military expedition and high risk work environments, such as firefighters.

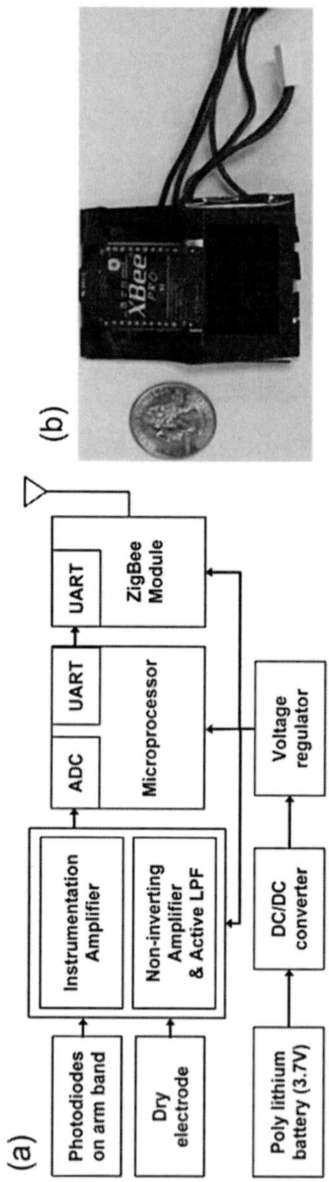

Figure 3. (**a**) Schematic of the sensor electronics module (SEM); (**b**) actual SEM used in e-bro system. Reproduced with permission from [47]. Copyright 2014, The Electrochemical Society.

3.2. E-Bra

The systems incorporated in the inner vest for men can also be integrated in an inner garment for women, such as a brassier. The various sensors listed in the previous section can be incorporated in the e-bra and the signals from the sensors brought to the eNanoflex [70] module through printed conductive traces or conductive threads. Figure 4 shows a picture of the e-bra, the eNanoflex module used for data acquisition and wireless transmission and the simple signal display interface that plots the data received from the eNanoflex module.

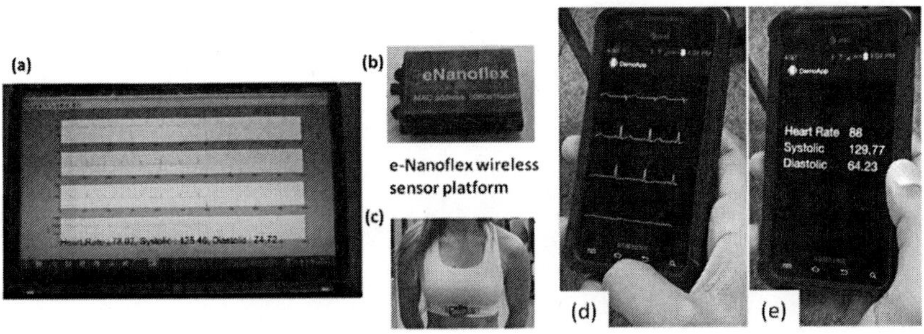

Figure 4. (**a**) Real-time data plotted on the laptop; (**b**) the eNanoflex module; (**c**) the e-bra worn by the test subject; (**d**) multi-channel signal smartphone display; (**e**) heart rate and blood pressure (systolic and diastolic) smartphone display.

3.2.1. Multichannel Data Acquired

The data acquired by the e-bro can be transmitted wirelessly to a PC. The data received by the PC was then filtered using an adaptive filter algorithm to minimize the effect of motion on the ECG signal baseline [71]. The data acquisition and adaptive filter were developed using MATLAB (Mathworks, Natick, MA, USA). However, the same can be achieved on a JAVA platform and can be deployed on a smartphone. Figure 4a,d shows the original three ECG signals, Lead I, Lead II and Lead III. It also plots the pulse waveform, the heart rate and the estimated systolic and diastolic blood pressure. Figure 4b shows the eNanoflex module used for data acquisition and wireless transmission. The derived pulse transit time (PTT) values are then used to estimate the systolic and diastolic blood pressure values (Figure 4e) based on the calibration equations previously obtained in [52]. Other sensor systems can be incorporated to develop wearable applications to monitor respiration, temperature and blood oxygen level.

Post processing of ECG can also calculate heart rate variability (HRV), which is a prognostic and diagnostic tool. HRV is described as the sequence formed by concatenating the difference in heart rate between consecutive beats (Figure 5a). It is calculated as the inverse of the difference in the intervals between consecutive R-peaks. The R-peak detection algorithm used for the calculation of the RR interval (RRI) was as given in [72]. The autoregressive (AR) power spectrum estimation technique was used to obtain the power spectrum density (PSD) plot of the RRI sequence. The characteristic LF and HF peaks were observed [73]. Figure 5b shows the AR PSD computed from the RRI series for the standing-up case. Figure 5c shows the same for supine ECG. AR PSDs in both figures show a classic shift in the power distribution between LF and HF components with respect to total power.

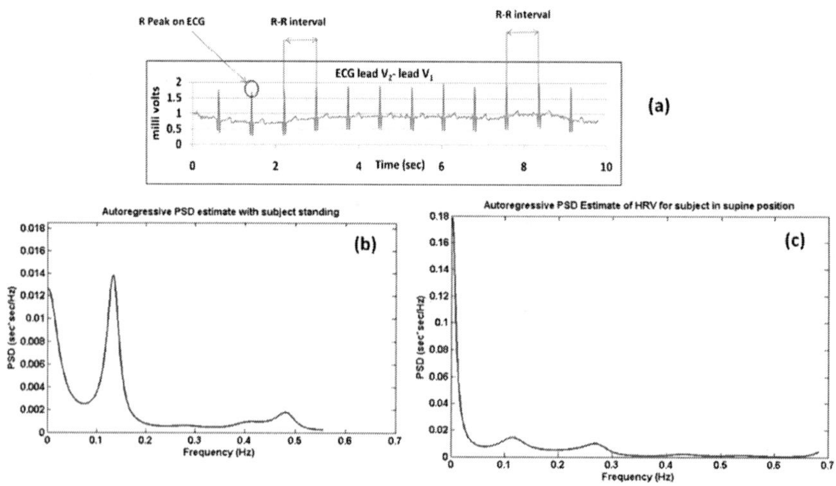

Figure 5. .(**a**) R-R interval calculations from ECG; (**b**) plot of the autoregressive (AR) power spectrum density (PSD) computed from the RR interval (RRI) series for the standing case; (**c**) plot of the AR PSD computed from the RRI series for the supine position.

Thus, these implementations of the e-bro and e-bra systems can be used for the tracking of chronic conditions related to autonomous nervous regulation of cardiac activity. Continuous multiple lead ECG monitoring can be used for the detection of T-wave inversion, which is indicative of a change in ventricular repolarization. Automated post processing of ECG by algorithm for the detection of T wave inversion can serve as an alarm system that will trigger a subroutine to initiate the ECG signal relay through a remote server to a doctor's office for diagnosis.

4. CONCLUSIONS

The need for wearable solutions for continuous cardiovascular monitoring from a clinical perspective was discussed in detail in this paper. Although research efforts have been underway over the past two decades in the field of dry electrodes for biopotential electrodes, recent advances in the use of nanomaterials and nanostructures in textiles have shown significant improvements in the performance of dry electrodes. In this review, we have identified the most recent advancements in the areas of nanotechnology in non-invasive cardiac biopotential monitoring systems that are wearable and textile based.

Furthermore, we have briefly discussed the wireless communication architectures, which also play a key role in the implementation of continuous wireless cardiovascular monitoring. Over the past few years, several research efforts have been underway to determine the communication architectures that could adequately support endeavors, like continuous remote patient monitoring and real-time analytics of biomedical signals, in a massively scalable fashion through cloud computing and the Internet of Things. Questions in this area are in areas such as data security, computational costs, sustainability in terms of energy and infrastructure costs. In this regard, we have briefly discussed the work done by our group, as well as alternative approaches to implementations.

Nonetheless, research on the communications technology perspective as well as the materials and sensors perspective have made significant strides in this direction over the past decade. The ultimate objective has been to design a truly personalized healthcare platform that can intelligently cater to each unique individual's healthcare needs. Recent trends in research are leading to this realization by tapping the benefits of nanotechnology, like the large surface area of dry electrodes with a small footprint.

ACKNOWLEDGMENTS

This research was conducted with the endowment, for smart textiles and healthcare, from the Global Institute for Nanotechnology in Engineering and

Medicine Inc. (700 Research Center Blvd., Fayetteville, AR 72701, USA). The authors would like to acknowledge Raj Mittra for suggesting that we communicate our work in this special issue for wearable electronics.

AUTHOR CONTRIBUTIONS

The research objectives were framed by Vijay Varadan and Robert E. Harbaugh. The introductory section and motivation for work were written by Prashanth Shyamkumar and Pratyush Rai, and reviewed by Vijay Varadan and Robert E. Harbaugh. The sections on sensors on textiles were written by Prashanth Shyamkumar and Pratyush Rai. Sections on systems implementation were written by Sechang Oh, Pratyush Rai and Mouli Ramasamy. Vijay Varadan and Robert E. Harbaugh provided guidance on the overall organization and preparation of the manuscript. All the authors read and approved the final manuscript.

CONFLICTS OF INTEREST

The authors declare no conflict of interest.

REFERENCES

1. Tao, X. *Smart Fibers, Fabrics and Clothing*; Woodhead Publishing: Cambridge, UK, 2001.

2. Zhang, X.; Tao, X. Smart textiles: Passive smart. *Textil Asia* **2001**, 45–49.

3. Zhang, X.; Tao, X. Smart textile: Active smart. *Textil Asia* **2001**, 49–52.

4. Zhang, X.; Tao, X. Smart textiles: Very smart. *Textile Asia* **2001**, 35–37.

5. Van Langenhove, L. *Smart Textiles for Medicine and Healthcare*, 1st ed.; Woodhead Publishing: Cambridge, UK, 2007.

6. Van Helleputte, N.; Konijnenburg, M.; Hyejung, K.; Pettine, J.; Dong-Woo, J.; Breeschoten, A.; Morgado, A.; Torfs, T.; de Groot, H.; van Hoof, C.; *et al.* A multi-parameter signal-acquisition SoC for connected personal health applications. In Proceedings of the 2014 IEEE International Solid-State Circuits Conference Digest of Technical Papers (ISSCC), San Francisco, CA, USA, 9–13 February 2014; pp. 314–315.

7. Searle, A.; Kirkup, L. A direct comparison of wet, dry and insulating bioelectric recording electrodes. *Physiol. Meas.* **2000**, *21*, 271.

8. California HealthCare Foundation. Snapshot Health Care Costs 101. Available online: http://www.chcf.org (accessed on 21 November 2013).

9. De la Maisonneuve, C.; Martins, J.O. *Public Spending on Health and Long-Term Care: A New Set of Projections*; OECD Publishing: Paris, France, 2013.

10. Thompson, P.D.; Franklin, B.A.; Balady, G.J.; Blair, S.N.; Corrado, D.; Estes, N.A., III; Fulton, J.E.; Gordon, N.F.; Haskell, W.L.; Link, M.S.; *et al.* Exercise and acute cardiovascular events placing the risks into perspective: A scientific statement from the American Heart Association Council on Nutrition, Physical Activity, and Metabolism and the Council on Clinical Cardiology. *Circulation* **2007**, *115*, 2358–2368.

11. Maron, B.J.; Doerer, J.J.; Haas, T.S.; Tierney, D.M.; Mueller, F.O. Sudden deaths in young competitive athletes: Analysis of 1866 deaths in the United States, 1980–2006. *Circulation* **2009**, *119*, 1085–1092.

12. Corrado, D.; Basso, C.; Pavei, A.; Michieli, P.; Schiavon, M.; Thiene, G. Trends in sudden cardiovascular death in young competitive athletes after implementation of a preparticipation screening program. *JAMA* **2006**, *296*, 1593–1601.

13. Myerburg, R.J.; Vetter, V.L. Electrocardiograms should be included in preparticipation screening of athletes.*Circulation* **2007**, *116*, 2616–2626.

14. Chaitman, B.R. An electrocardiogram should not be included in routine preparticipation screening of young athletes.*Circulation* **2007**, *116*, 2610–2614.

15. Faber, L.; van Buuren, F. Athlete screening for occult cardiac disease: No risk, no fun? *J. Am. Coll. Cardiol.* **2008**, *51*, 1040–1041.

16. Maron, B.J. National electrocardiography screening for competitive athletes: Feasible in the United States? *Ann. Int. Med.* **2010**, *152*, 324–326.

17. Corrado, D.; Pelliccia, A.; Bjornstad, H.H.; Vanhees, L.; Biffi, A.; Borjesson, M.; Panhuyzen-Goedkoop, N.; Deligiannis, A.; Solberg, E.; Dugmore, D.; *et al.* Cardiovascular preparticipation screening of young competitive athletes for prevention of sudden death: Proposal for a common European protocol. Consensus Statement of the Study Group of Sport Cardiology of the Working Group of Cardiac Rehabilitation and Exercise Physiology and the Working Group of Myocardial and Pericardial Diseases of the European Society of Cardiology. *Eur. Heart J.* **2005**, *26*, 516–524.

18. Maron, B.J.; Thompson, P.D.; Ackerman, M.J.; Balady, G.; Berger, S.; Cohen, D.; Dimeff, R.; Douglas, P.S.; Glover, D.W.; Hutter, A.M., Jr.; *et al.* Recommendations and considerations related to preparticipation screening for cardiovascular abnormalities in competitive athletes: 2007 update: A scientific statement from the American Heart Association Council on Nutrition, PhysicalActivity, and Metabolism: Endorsed by the American College of Cardiology Foundation. *Circulation* **2007**, *115*, 1643–1655.

19. Pelliccia, A.; Maron, B.J.; Culasso, F.; di Paolo, F.M.; Spataro, A.; Biffi, A.; Caselli, G.; Piovano, P. Clinical significance of abnormal electrocardiographic patterns in trained athletes. *Circulation* **2000**, *102*, 278–284.

20. Pelliccia, A.; di Paolo, F.M.; Quattrini, F.M.; Basso, C.; Culasso, F.; Popoli, G.; de Luca, R.; Spataro, A.; Biffi, A.; Thiene, G.; *et al.* Outcomes in athletes with marked ECG repolarization abnormalities. *N. Engl. J. Med.* **2008**, *358*, 152–161.

21. Pelliccia, A. Differences in cardiac remodeling associated with raceimplications for pre-participation screening and the unfavorable situation of black athletes. *J. Am. Coll. Cardiol.* **2008**, *51*, 2263–2265.

22. Uberoi, A.; Stein, R.; Perez, M.V.; Freeman, J.; Wheeler, M.; Dewey, F.; Peidro, R.; Hadley, D.; Drezner, J.; Sharma, S.; *et al.* Interpretation of electrocardiogram of young athletes. *Circulation* **2011**, *124*, 746–757.

23. Drezner, J.A.; Courson, R.W.; Roberts, W.O.; Mosesso, V.N., Jr.; Link, M.S.; Maron, B.J. Inter-association task force recommendations on emergency preparedness and management of sudden cardiac arrest in high school and college athletic programs: A consensus statement. *J. Athl. Train.* **2007**, *42*, 143–158.

24. Rubart, M.; Zipes, D.P. Mechanisms of sudden cardiac death. *J. Clin. Invest.* **2005**, *115*, 2305–2515.

25. Eckart, R.E.; Scoville, S.L.; Campbell, C.L.; Shry, E.A.; Stajduhar, K.C.; Potter, R.N.; Pearse, L.A.; Virmani, R. Sudden death in young adults: A 25-year review of autopsies in military recruits. *Ann. Int. Med.* **2004**, *141*, 829–834.

26. Eckart, R.E.; Shry, E.A.; Burke, A.P.; McNear, J.A.; Appel, D.A.; Castillo-Rojas, L.M.; Avedissian, L.; Pearse, L.A.; Potter, R.N.; Tremaine, L.; *et al*. Sudden death in young adults—An autopsy-based series of a population undergoing active surveillance. *J. Am. Coll. Cardiol.* **2011**, *58*, 1254–1261.

27. Park, S.; Gopalsamy, C.; Rajamanickam, R.; Jayaraman, S. The wearable motherboard™: An information infrastructure or sensate liner for medical applications. *Stud. Health Technol. Inform.* **1999**, *62*, 252–258.

28. Park, S.; Jayaraman, S. Wearable biomedical systems: Research to reality. In Proceedings of the IEEE International Conference on Portable Information Devices PORTABLE07, Orlando, FL, USA, 25–29 May 2007; pp. 1–7.

29. Coosemans, J.; Hermans, B.; Peurs, R. Intergrating wireless ECG in textiles. *Sens. Actuators A* **2006**, *130–131*, 48–53.

30. Lee, Y.D.; Chung, W.Y. Wireless sensor network based wearable smart shirt for ubiquitous health and activity monitoring. *Sens. Actuators B* **2009**, *140*, 390–395.

31. Alzaidi, A.; Zhang, L.; Bajwa, H. Smart Textiles Based Wireless ECG System. In Proceedings of the 2012 IEEE Long Island Systems, Applications and Technology Conference (LISAT), Farmingdale, NY, USA, 4 May 2012; pp. 1–5.

32. Smart T-shirt by GOW Trainer. Available online: http://www.gowtrainer.com (accessed on 21 November 2013).

33. Wearable Wellness System from Smartex s.r.l. Pisa Italy. Available online: http://www.smartex.it (accessed on 22 November 2013).

34. Rattfalt, L.; Chedid, M.; Hult, P.; Linden, M.; Ask, P. Electrical properties of textile electrodes. In Proceedings of the 29th Annual International Conference of the IEEE Engineering in Medicine and

Biology Society, 2007. EMBS 2007, Lyon, France, 22–26 August 2007; pp. 5735–5738.

35. Fernandes, M.S.; Lee, K.S.; Ram, R.J.; Correia, J.H.; Mendes, P.M. Flexible PDMS-based dry electrodes for electro-optic acquisition of ECG signals in wearable devices. In Proceedings of the 2010 Annual International Conference of the IEEE Engineering in Medicine and Biology Society (EMBC), Buenos Aires, Argentina, 31 August–4 September 2010; pp. 3503–3506.

36. Ruffini, G.; Dunne, S.; Farres, E.; Cester, I.; Watts, P.C.P.; Ravi, S.; Silva, P.; Grau, C.; Fuentemilla, L.; Marco-Pallares, J.;*et al*. ENOBIO dry electrophysiology electrode; first human trial plus wireless electrode system. In Proceedings of 29th Annual International Conference of the IEEE Engineering in Medicine and Biology Society, 2007, EMBS 2007, Lyon, France, 22–26 August 2007; pp. 6689–6693.

37. Jung, H.; Moon, J.; Baek, D.; Lee, J.H.; Choi, Y.Y.; Hong, J.S.; Lee, S.H. CNT/PDMS composite flexible dry electrodesfor long-term ECG monitoring. *IEEE Trans. Biomed. Eng.* **2012**, *59*, 1472–1479.

38. Baek, J.; An, J.-H.; Choi, J.-M.; Park, K.-S.; Lee, S.-H. Flexible polymeric dry electrodes for the long-term monitoring of ECG. *Sens. Actuators A* **2008**, *143*, 423–429.

39. Varadan, V.K.; Oh, S.; Kwon, H.; Hankins, P. Wireless point-of-care diagnosis for sleep disorder with dry nanowire electrodes. *J. Nanotechnol. Eng. Med.* **2010**, *1*, 031012.

40. Fiedler, P.; Griebel, S.; Fonseca, C.; Vaz, F.; Zentner, L.; Zanow, F.; Haueisen, J. Novel Ti/TiN dry electrodes and Ag/AgCl: A direct comparison in multichannel EEG. In Proceedings of the 5th European Conference of the International Federation for Medical and Biological Engineering, Budapest, Hungary, 14–18 September 2012; pp. 1011–1014.

41. Dias, N.S.; Carmo, J.P.; Ferreira da Silva, A.; Mendes, P.M.; Correia, J.H. New dry electrodes based on iridium oxide (IrO) for non-invasive biopotential recordings and stimulation. *Sens. Actuators A* **2010**, *164*, 28–34.

42. Chiou, J.-C.; Ko, L.-W.; Lin, C.-T.; Hong, C.-T.; Jung, T.-P.; Liang, S.-F.; Jeng, J.-F. Using novel MEMS EEG sensors in detecting drowsiness application. In Proceedings of the 2006 IEEE Biomedical Circuits and Systems Conference (BioCAS 2006), London, UK, 29 November–1 December 2006; pp. 33–36.

43. Chi, Y.M.; Tzyy-Ping, J.; Cauwenberghs, G. Dry-contact and noncontact biopotential electrodes: Methodological review. *IEEE Rev. Biomed. Eng.* **2010**, *3*, 106–119.

44. Mestrovic, M.A.; Helmer, R.J.N.; Kyratzis, L.; Kumar, D. Preliminary study of dry knitted fabric electrodes for physiological monitoring. In Proceedings of the 3rd International Conference on Intelligent SensorsSensor Networks and Information, 2007, ISSNIP 2007, Melbourne, Australia, 3–6 December 2007; pp. 601–606.

45. Shim, B.S.; Chen, W.; Doty, C.; Xu, C.; Kotov, N.A. Smart electronic yarns and wearable fabrics for human biomonitoring made by carbon nanotube coating with polyelectrolytes. *Nano Lett.* **2008**, *8*, 4151–4157.

46. InOh, T.; Yoon, S.; Kim, T.E.; Wi, H.; Kim, K.J.; Woo, E.J.; Sadleir, R.J. Nanofiber web textile dry electrodes for long-term biopotential recording. *IEEE Trans. Biomed. Circuits Syst.* **2013**, *7*, 204–211.

47. Rai, P.; Oh, S.; Shyamkumar, P.; Ramasamy, M.; Harbaugh, R.E.; Varadan, V.K. Nano-bio-textile sensors with mobile wireless platform for wearable health monitoring of neurological and cardiovascular disorders. *J. Electrochem. Soc.***2014**, *161*, B3116–B3150.

48. Karaguzel, B.; Merritt, C.R.; Kang, T.; Wilson, J.M.; Nagle, H.T.; Grant, E.; Pourdeyhimi, B. Utility of nonwovens in the production of integrated electrical circuits via printing conductive inks. *J. Textile Inst.* **2008**, *99*, 37–45.

49. Araki, T.; Nogi, M.; Suganuma, K.; Kogure, M.; Kirihara, O. Printable and stretchable conductive wirings comprising silver flakes and elastomers. *IEEE Electron Device Lett.* **2011**, *32*, 1424–1426.

50. Romaguera, V.S.; Madec, M.B.; Yeates, S.G. Inkjet printing of conductive polymers for smart textiles and flexible electronics. *Mater. Res. Soc. Symp. Proc.* **2009**, *1192*, 26–31.

51. Rai, P.; Lee, J.; Mathur, G.N.; Varadan, V.K. Carbon nanotubes polymer nanoparticle inks for healthcare textile. *Proc. SPIE* **2012**, *8548*.

52. Rai, P.; Kumar, P.S.; Oh, S.; Kwon, H.; Mathur, G.N.; Varadan, V.K.; Agarwal, M.P. Smart healthcare textile sensor system for unhindered-pervasive health monitoring. *Proc. SPIE* **2012**, *8344*.

53. Kumar, P.S.; Rai, P.; Oh, S.; Kwon, H.; Varadan, V.K. Nanocomposite electrodes for smartphone enabled healthcare garments: E-bra and smart vest. *Proc. SPIE* **2012**, *8548*.

54. Zhang, Y.-T.; Poon, C.C.Y.; Chan, C.-H.; Tsang, M.W.W.; Wu, K.F. A health-shirt using e-textile materials for the continuous and cuffless monitoring of arterial blood pressure. In Proceedings of 3rd IEEE/EMBS International Summer School on Medical Devices and Biosensors, 2006, Cambridge, MA, USA, 4–6 September 2006; pp. 86–89.

55. Ghoshdastider, U.; Lange, C.; Viga, R.; Grabmaier, A. A modular and wireless exg signal acquisition system with a dense array of dry electrodes. In Proceedings of 2012 IEEE Sensors, Taipei, Taiwan, 28–31 October 2012; pp. 1–4.

56. Bifulco, P.; Cesarelli, M.; Fratini, A.; Ruffo, M.; Pasquariello, G.; Gargiulo, G. A wearable device for recording of biopotentials and body movements. In Proceedings of 2011 IEEE International Workshop on Medical Measurements and Applications Proceedings (MeMeA), Bari, Italy, 30–31 May 2011; pp. 469–472.

57. Lopez, G.; Custodio, V.; Moreno, J.I. LOBIN: E-Textile and Wireless-Sensor-network-Based Platform for Healthcare Monitoring in Future Hospital Environment. *IEEE Trans. Inf. Technol. Biomed.* **2010**, *14*, 1446–1458.

58. Coyle, S.; Lau, K.-T.; Moyna, N.; O'Gorman, D.; Diamond, D.; Di Francesco, F.; Costanzo, D.; Salvo, P.; Trivella, M.G.; de Rossi, D.E.; *et al.* BIOTEX—Biosensing Textiles for Personalised Healthcare Management. *IEEE Trans. Inf. Technol. Biomed.* **2010**, *14*, 364–370.

59. Hung, K.; Lee, C.C.; Chan, W.M.; Sheung-On, C.; Kwok, P. Development of a wearable system integrated with novel biomedical sensors for ubiquitous healthcare. In Proceedings of the 2012 Annual International Conference of the IEEE Engineering in Medicine and Biology Society (EMBC), San Diego, CA, USA, 28 August–1 September 2012; pp. 5802–5805.

60. Axisa, F.; Scmitt, P.M.; Gehin, C.; Delhomme, G.; McAdams, E.; Dittmar, A. Flexible Technologies and Smart Clothing for Citizen Medicine, Home Healthcare, and Disease Prevention. *IEEE Trans. Inf. Technol. Biomed.* **2005**, *9*, 325–336.

61. Steltenkamp, S.; Becher, K.; Doerge, T.; Ruff, R.; Hoffmann, K.-P. Electrode structures for acquisition and neural stimulation controlling the cardiovascular system. In Proceedings of the Annual International Conference of the IEEE Engineering in Medicine and Biology Society, 2009, EMBC 2009, Minneapolis, MN, USA, 3–6 September 2009; pp. 5478–5481.

62. Chamadiya, B.; Mankodiya, K.; Wagner, M.; Nasreddine, R.B.; Hofmann, U.G. Non-contact, non-obtrusive electrocardiography in clinical environements. In Proceedings of the the 2011 5th International Conference on Pervasive Computing Technologies for Healthcare (PervasiveHealth), Dublin, Ireland, 23–26 May 2011; pp. 101–106.

63. Beckmann, L.; Jacob, M.; Antink, C.H.; Cordes, A.; Pikkemaat, R.; Jungbecker, N.; Gries, T.; Leonhardt, S. Portable bioimpedance spectroscopy device and textile electrodes for mobile monitoring applications. *Int. Conf. Electr. Bioimpedance* **2010**, *224*, 1–4.

64. Lagunar, C.M.; Barca, C.C.; Padrón, A.M.Q.; Planes, X.; Wattenberg, F.S.; López, C.A.; Martín-Fernández, M.; Hernández, N.M.; Oliveras, E.C.; Herranz, J.C. Ubiquitous Tele-monitoring Kit (UTK): Measuring physiological signals anywhere at anytime. In *Ambient Assisted Living and Home Care*; Springer: Berlin/Heidelberg, Germany, 2012; Volume 7657, pp. 183–191.

65. Doukas, C.; Maglogiannis, I. Managing wearable sensor data through cloud computing. In Proceedings of the 2011 IEEE Third International Conference on Cloud Computing Technology and Science (CloudCom), Athens, Greece, 29 November–1 December 2011; pp. 440–445.

66. Yoon, H.; Deshpande, D.C.; Ramachandran, V.; Varadan, V.K. Aligned nanowires growth using lithography-assisted bonding of a polycarbonate template for neural probe electrodes. *Nanotechnology* **2008**, *19*, 025304.

67. Reneker, D.H.; Chun, I. Nanometer diameter fibers of polymer, produced by electrospining. *Nanotechnology* **1996**, *7*, 216–223.

68. Laukhina, E.; Pfattner, R.; Mas-Torrent, M.; Rovira, C.; Veciana, J.; Laukhin, V. Flexible film-based sensors structured with a high piezoresistive organic molecular conductor as an active component. In Proceedings of the 2010 First International Conference on Sensor Device Technologies and Applications (SENSORDEVICES), Venice, Italy, 18–25 July 2010; pp. 4–9.

69. Castellanos, J.; Navas-Gonzalez, R.; Ochoteco, E.; Vidal-Verdu, F. Evaluation of a low cost piezorresistive material for high resolution tactile sensors. In Proceedings of the 2008 2nd European Conference & Exhibition on Integration Issues of Miniaturized Systems - MOMS, MOEMS, ICS and Electronic Components (SSI), Barcelona, Spain, 9–10 April 2008; pp. 1–3.

70. Varadan, V.K.; Kumar, P.S.; Oh, S.; Kwon, H.; Rai, P.; Banerjee, N.; Harbaugh, R.E. E-nanoflex sensor systems: Smartphone-based roaming health monitor. *J. Nanotechnol. Eng. Med.* **2011**, *2*, 011016.

71. Kwon, H.; Oh, S.; Varadan, V.K. Motion artifact removal algorithm by ICA for e-bra: A women ECG measurement system. *Proc. SPIE* **2013**, *8691.*

72. Pan, J.; Tompkins, W.J. A real-time QRS detection algorithm. *IEEE Trans. Biomed. Eng.* **1985**, *32*, 230–236.

73. Malik, M.; Bigger, J.T.; Camm, A.J.; Kleiger, R.E.; Malliani, A.; Moss, A.J.; Schwartz, P.J. Task force of the european society of cardiology the north american society of pacing electrophysiology. Heart rate variability: Standards of measurement, physiological interpretation, and clinical use. *Circulation* **1996**, *93*, 1043–1065.

CHAPTER 8

RFID in textile and clothing manufacturing: technology and challenges

Rajkishore Nayak, Amanpreet Singh, Rajiv Padhye* and Lijing Wang

School of Fashion and Textiles, RMIT University, Melbourne, Australia

ABSTRACT

In the field of textile and clothing, radio frequency identification (RFID), which is one of the most promising technological innovations, is used in manufacturing, inventory control, warehousing, distribution, logistics, automatic object tracking and supply chain management. Various retailers and manufacturers (of clothing as well as consumer goods) such as CVS, Tesco, Prada, Benetten, Wal-mart and Procter & Gamble, are now implementing the technology and exploring the impact of the technology on their business. RFID technologies may improve the potential benefits of supply chain management through reduction of inventory losses, increase of the efficiency and speed of processes and improvement of information accuracy. The basic of success lies in understanding the technology and other features to minimize the potential problems. Although the technology existed for several years, the technological challenges and cost issues are the major hurdles for the widespread use of RFID. In this paper, the authors have addressed the technology of RFID and various applications related to inventory management, production control, retail management, brand segregation etc. in textile and clothing industry. In addition, the disadvantages, challenges and future directions of RFID technology have also been highlighted.

KEYWORDS

RFID, Fashion, Textiles, Supply chain management, Inventory, Retail

1. INTRODUCTION

The rapid pace at which the technological innovations are being introduced in the world poses a potential challenge to the retailer, supplier, and enterprises. In the field of Information Technology (IT) there is a rapid growth in the last 30 years (Want 2006; Landt 2005). One of the most promising technological innovations in IT is radio frequency identification (RFID) (Dutta et al. 2007; Whitaker et al. 2007; Bottani et al. 2009). The RFID technology was evolved in 1945 as an espionage tool invented by Leon Theremin for the Soviet Government (Nikitin et al. 2013, Tedjini et al. 2012). At that time it was mainly used by the military. The progress in microchip design, antenna technology and radio spread spectrum pushed it into various applications like supply chain management, retail, automatic toll collection by tunnel companies, animal tracking, ski lift access, tracking library books, theft prevention, vehicle immobilizer systems, railway rolling stock identification, movement tracking, security, healthcare, printing, textiles and clothing (Weinstein 2005; Liu and Miao 2006; Rao et al. 2005; Wu et al. 2009; Tan 2008). RFID can make the companies more competitive by changing the related processes in supply chain, manufacturing and retailing.

The RFID and bar coding systems resemble in the concept of identifying different objects (White et al. 2007; Dos Santos and Smith 2008). However, RFID enhances data processing and works as a complementary to the existing technologies. RFID is the advanced system over barcodes in terms of non-optical proximity communication, information density and two-way communication ability (Roberts 2006; Nath et al. 2006; Miles et al. 2008). The recent developments in microelectronics and data processing have enabled the use of less expensive and smaller components with improved efficiency of RFID. Hence, the application of RFID in various areas particularly in supply chain has grown from $1 billion in 2003 to $20 billion in 2013 (Bagchi et al. 2007). The recent research works on RFID focus on inventory management, logistics, transportation, manufacturing, asset tracking and object location (Piramuthu et al. 2014; Sun 2012; García et al. 2007; Nayak and Padhye 2015a).

World trade organisation (WTO) is continuously thriving to facilitate international trade by reducing barriers which assists the manufacturing process of fashion products to low cost countries. Christopher characterizes the fashion industry by wider collection/more choices, seasonally impacted, variable and short life-cycles, complex supply chain and distribution activities (Christopher et al. 2004). Profitability in fashion industries are going down due to the competitive nature of business, thus requiring better coordination between all the partners in supply chain. Fashion companies dominating the market at some stage have lost their market share to the emerging companies like Zara, H&M and Benetton as the business strategy of these companies' involves more inventory turns and they offer wider range of assortments giving rise a concept of "Fast Fashion". All these changes in fashion industry need to capture new technologies and using them wisely, which can reduce lead time, cost and most importantly efficiently manage the inventory. Among the new technologies, RFID is of particular interest as it has capability of providing solutions to some of these issues (Moon and Ngai 2008).

Barcode systems, although used for product information, inventory control, have some drawbacks as compared to RFID (Delen et al. 2007; Kärkkäinen 2003). The amount of information stored in a barcode is very less as compared to RFID. RFID can store up to 1,000 bytes of data. An RFID is specific to each item, whereas the barcode is not. Barcode needs human interaction for proper operation. It requires time-of-sight access to an optical scanner for the product related information. The barcode is to be replaced if the information it contains needs modification, but in RFID it can be modified at stages of the supply chain by the interaction between the microchip and the reader software. The barcode system is less accurate as compared to RFID.

While the use of RFID systems can improve the productivity and efficiency, they also expose new threats to the security and privacy of individual organisations (Phillips et al.2005; Bogdanov et al. 2008; Ayoade 2007). The attackers can easily use duplicate tags to provide wrong information or can break the security algorithm or can modify the message. Hence, research and development should focus on resolving these issues. The other major problem associated with RFID is the lack of standardization as there are varieties of RFID tags (Michael and McCathie 2005; Goodrum et al. 2006; Bunduchi et al. 2011). The effort of governing bodies to develop various standards related to their use, manufacture and technical specification can help to resolve the issues related to standardization. In addition, the technological developments and innovations in information technology will also assist in this direction to achieve wide adoptability of RFID tags.

Although several RFID research publications focused on various applications of RFID, there is a dearth of review of the literature on the application of RFID in textile and clothing manufacturing and retailing. Hence, an attempt has been made in this paper to review various applications of RFID in the textile and clothing, manufacture and distribution. This paper also covers the technology of RFID, major challenges faced by the technology and the future directions.

2. RFID SYSTEMS

2.1 Technology of RFID system

Radio frequency identification is an automatic identification method consisting of several components such as tags, tag readers, edge servers, middleware, and application software (Nayak et al. 2007; Rao et al. 2005; Chao et al. 2007). Among these the three important components are RFID tag (also known as transponder), RFID reader (also known as transceiver or interrogator) and software for data processing. An RFID tag is a small object that can be attached to or embedded into a product, animal, or person. It consists of a tiny chip where the data is stored and an antenna to enable it to receive and respond to radio-frequency queries from an RFID transceiver (See Figure 1). The tags contain electronic product code (EPC) and the information related to the product like the name of the company, batch and year of manufacturing, price etc. (Sarma et al. 2001; Myny et al.2010).

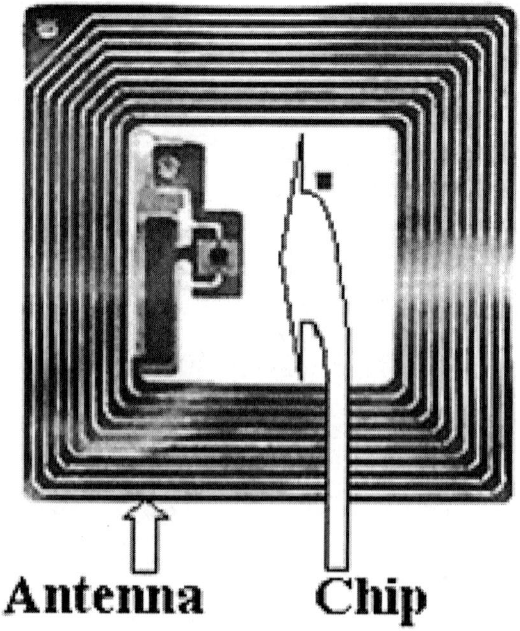

Figure 1. A RFID tag with chip and antenna.

There are four main frequency bands for RFID tags commonly in use. They are categorized by their radio frequency: low frequency tags (125 or 134.2 kHz), high frequency tags (13.56 MHz), UHF tags (868–956 MHz), and microwave tags (2.45 or 5.8 GHz). RFID tags can be either active, semi-passive (i.e. semi-active) or passive (Nayak et al. 2007; Tesoriero et al. 2008; Rao et al. 2005; Costin et al. 2012). Various applications of RFID according to the frequency band are discussed in Table 1.

2.2 Types of RFID systems

Passive RFID tags have no internal power supply. The minute electrical current induced in the antenna by the incoming radio frequency signal provides just enough power for the integrated circuit (IC) in the tag to power up and transmit a response (Bouet and Dos Santos 2008). Most passive tags signal by backscattering the carrier signal from the reader. This means that the aerial (antenna) has to be designed to both collect power from the incoming signal and also to transmit the outbound backscatter signal. The tag chip can contain nonvolatile EEPROM (Electrically Erasable Programmable Read-Only Memory) for storing data (Barnett and Liu 2008). Lack of an onboard power supply means that the device can be quite small: commercially available products exist that can be embedded under the skin. As of 2005, the smallest such devices commercially available measured 0.4 mm × 0.4 mm, and is thinner than a sheet of paper; such devices are practically invisible. Passive tags have practical read distances ranging from about 2 mm up to about few metres depending on the chosen radio frequency.

Semi-passive RFID tags are very similar to passive tags except for the addition of a small battery (Zhang and Amin 2006). This battery allows the tag IC to be constantly powered (Li et al. 2009). This removes the need for the aerial to be designed to collect power from the incoming signal. Aerials can therefore be optimised for the backscattering signal. Semi-passive RFID tags are faster in response as compared to passive tags. *Active RFID*tags or beacons, on the other hand, have an internal power source which is used to power any ICs and generate the outgoing signal (Ni et al. 2004; Jin and Lu 2006). They may have longer range and larger memories than the passive tags, as well as the ability to store additional information sent by the transceiver. At present, the smallest active tags are about the size of a coin. Many active tags have practical ranges of tens of metres, and a battery life of up to 10 years. As passive tags are cheaper to manufacture and have no battery, the majority of RFID tags in existence are of this variety.

Table 1. RFID frequency bands and various applications

Frequency band	Features	Applications
100–500 kHz (low)	Short to medium read range; inexpensive; low reading speed	Access control; animal identification; inventory control; car immobilizer
10–15 MHz (intermediate)	Short to medium read range; potentially inexpensive; medium reading speed	Access control; smart cards; library control
850–950 MHz, 2.4–5.8 GHz (high)	Long read range; high reading speed; line of sight required; expensive	Railway vehicle monitoring; toll collection systems; pallet and container tracking; vehicle tracking

Data taken from radio frequency identification—a basic primer, Association for Automatic Identification and Mobility, August 2001, http://www.aimglobal.org and from Wikipedia Encyclopedia, http://en.wikipedia.org/wiki/RFID.

2.3 Working of an RFID systems

The reader communicates with a tag, which contains digital information (Figure 2). A collection of wired and wireless network constitutes the communication infrastructure that carries out a series of information transfer actions to deliver the data stored in a tag to the reader. Readers are just like the barcode sensors, which broadcast a radio signal through the antenna to the tag. The tag then responds to the radio wave, and the data can be read from the chip of the tag. The tags can consist of read only memory (ROM), volatile read/write random access memory (RAM) or write once/read many memory (WORM) for data storage (Jin and Lu 2006). ROM is used to store security data, whereas RAM is used for storing the data during transponder interrogation and response (Dimitriou 2005). The data consists of: (a) a unique identifier, (b) an operating system, (c) data storage (volatile/non-volatile), and (d) an EPC.

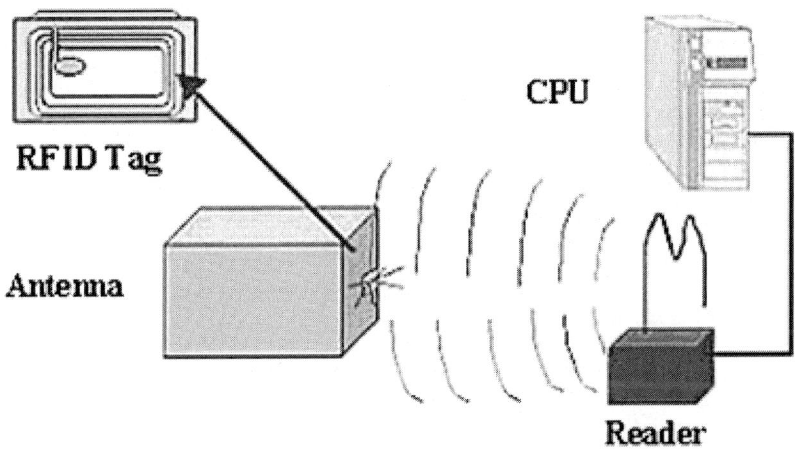

Figure 2. Working of RFID system.

Simultaneously data from multiple tags can be studied. Readers also decode the signal provided by the tags and transmit to the CPU. Suitable Software receives and interprets the information collected form the tags and stores it. Simultaneously multiple data can be collected and stored with much accuracy as compared to the existing barcode systems.

2.4 Applications of RFID in fashion industry

In the field of Textile and Apparel the RFID is used in manufacturing, inventory control, warehousing, distribution, logistics, automatic object tracking and supply chain management (Legnani et al. 2011; Liu et al. 2010; Moon and Ngai 2008; Gimpel et al.2004). For example, the finished garments, different pattern pieces and accessories can be traced and the progress of the production process can be monitored. In processing and weaving, the fabric lots can be traced easily. In spinning mills, the bales of cotton and the yarns can be traced easily. The mixing of different yarn lots which is a major problem in spinning mills can be avoided.

A range of problems unique to fashion industry can be solved by installing the RFID system in various processes involved in fashion supply chain. Fashion industry can be divided into four sectors such as manufacturing, overseas transportation, distribution and retailing. In manufacturing RFID can be used to avoid product and component mixing, and mixing of different accessories. In overseas transportation RFID can be used to track and trace the consignment before and after Free on Board, if product is outsourced. RFID primarily can solve the issue of sorting and tracking the product very quickly, thus reducing the lead time during distribution (Moon and Ngai 2008). In majority of the retail shops, the space is a constraint, which needs to be used efficiently. RFID is used to manage the stock to identify the item quickly and correctly so as to avoid “search regret”, and thus help to solve the problem (Reynolds et al. 2006).

2.5 Inventory management

The use of RFID system assists in better inventory management and enhances customer shopping experience (Hardgrave et al. 2009; De Kok et al. 2008). The RFID system can indicate the type of clothing selling faster or slower, which provide customized advertisement on picked clothes according to the RFID tags. The RFID use also helps in reducing labour. For example, American Apparel applied RFID system in eight of their stores, which saved about 60–80 h per week in labor and reduced out-of-stock products owing to unawareness. Similarly, the Japanese apparel manufacturer ‘Sankei’ is using a RFID system in manufacturing to track clothes during the manufacturing process in addition to the online sales (Wu et al. 2009).

A successful case study has been conducted by Kaufhaulf departmental store in Europe on the impact of RFID on their processes and supply chain (Loebbecke and Huyskens 2008). They have concluded that RFID can be used in warehousing to track the goods and assist in cross docking operation as the goods received are not segregated in the warehouse but stored as they are received from the manufacturer. In the distribution center's receivable area RFID readers can be implemented to track the goods as they come in. The received goods are registered and the data stored on RFID transponders can be read for each good though received in pallets or individual fashion item on hanger. In this the task of counting hangers, items and pallets can be eliminated and quantity can be controlled up to 100% satisfaction level.

2.6 Production control

Inside the stores of the production department the stock level can be quickly monitored which helps in a real-time and efficient inventory management (Wang et al. 2006; Lee et al. 2013; Jihui et al. 2011). The location and movement of the components, semi-finished and the finished products can be detected by the RFID system, which helps in production monitoring and control. The performance of departments and the individuals can be evaluated by the rate of movement of the garment components from the departments or individuals. This helps in improving the productivity and the quality. In the packing floor the mixing of different styles and sizes can be avoided by RFID tags. Also the number of pieces in packed cartons can be counted without opening, which saves the time and labour.

2.7 Retail management

It can be used in retail to monitor and control the floor level out-of-stocks (OOS) (Roussos2006; Chen and Pfleuger 2008). It is mostly required where there is high product display density, low staffing level and chances of mishandling is very high. The stock level of the items is properly maintained and the items can be grouped according to their demand. Similarly, while receiving the goods at the store the accuracy of the deliveries can be ensured quickly with less labour. This can improve the inventory accuracy and automatic replenishment of stock. Mostly the retailers verify the accuracy and integrity of the shipment by open-box audit prior to stocking or storing of goods. But now it is possible to read all the items packed inside the carton without opening, which saves time and labour cost. So the RFID provides a new horizon to the retailers in receiving of materials, which eliminates the invoice disputes, mixing of different quality products etc.

The RFID has been used by number of major retailers in Europe and USA. Walmart is among the early adopters of RFID throughout their supply chain and the suppliers working with Walmart had been under pressure to implement RFID in their supply chain component as well. The reason of implementing RFID is line with Walmart's corporate strategy of being cost competitiveness and faster response (Vowels 2006). After successfully implementation and use of RFID by Walmart, other big retailers like Marks and Spencer and Tesco in UK are adopting it, Coles group in Australia have given a thought of embracing this technology. Fashion industry is always regarded as the late adopter of new technology than others, but in adopting RFID it is not lagging behind.

The theft of garments from the fitting room can be prevented by mounting a small reader at the entry point of the room. The tag information of the garment is captured when the customer takes the garment to the fitting room. The items that are taken to the fitting room but not come out are reported as potential loss items. So the RFID can be used to identify the missing merchandise and prosecution of the shoplifters (Rekik et al. 2008; Jaselskis et al. 1995). In addition, the dislocation of the clothes is automatically detected and customers can be given advice on the fitting of the clothes they are trying on. The later task can be achieved by the use of smart try-out cabins and smart mirrors. The smart mirror can advise on additional clothing and fashion accessories (Roberti 2006). Companies such as Marks and Spencer, Levi Strauss & Co. are installing RFID to acquire data on the amount of stocks every day so that the stock can be replenished. However, significant difference was observed between the data received from the RFID systems and the actual inventory levels.

The store merchandise recovery and replenishment process presently used is completely manual, unsystematic and ineffective. Also it requires a very long time and high labour force and some popular items are not replenished for weeks. This can be overcome by RFID technology. The tags help to estimate the stock level frequently and there is an in-stock position improvement up to 30%. This has a positive impact on sales and customer satisfaction. It prevents the fashion or the seasonal merchandise to lie undetected in the back room and to be sold at significant markdowns thus improving the profit margin.

Also at the point of sales the RFID tag can be read quickly and it avoids the physical handling of the product as in barcode system. The RFID tags can be placed at the side seams or hem or in a paper cartoon label without affecting the aesthetic appeal of the garment. The number of items sold is automatically adjusted from the amount of stock. Hence, the customers can be informed about the status/stock of merchandise by interactive display screens. The use of RFID tags has helped to reduce the out of stock rate of Wal-mart by 16% and RFID enabled stores were 63% more effective in replenishing out of-stocks than control stores (Age 2005).

2.8 RFID for brand segregation

It will be quite beneficial to adopt RFID system for departmental stores, which stocks a range of products and brands under one roof (Loebbecke and Huyskens 2008). The different brands can be controlled in terms of their inventory and location quite easily using RFID whereas it would be quite challenging task to segregate the brands in case they are mixed. RFID also reduces the human error as this technology is least human dependent when monitoring the inventory level, brand identification and product authenticity.

Radio frequency identification can also be used to avoid illegal product alteration to grey market and increasing product security and to give information about shopping behavior of customer at retail. Having all these issues solved, textile and fashion supply chain can be made more streamlined and efficient in terms of cost and performance.

2.9 RFID used in fitting room

Radio frequency identification technology is used in the fitting room in a new approach by a German departmental store "Galeria Kaufhof in Essen", (part of the Metro retailing group). Automatic suggestions are provided to men buying clothes in the store when they go to a dressing room to try a suit. A 'smart mirror' informs them what kind of suit or accessories will be appropriate. An RFID reader on the smart mirror in the fitting room determines which clothing has been brought into the room from the RFID tag attached to the apparel. The appropriate accessories are then displayed on a suitable interface. This system is used in combination with 'smart shelves', which can read what merchandise is currently in stock, so that customers can be shown various styles and sizes that are available.

2.10 RFID tags for care labelling

The recent care labels are printed or woven labels storing a limited information (Nayak and Padhye 2014, 2015b). The electronic labels use RFID tags for storing information electronically on a garment. The same RFID tag used for containing the product details at the point of sale cannot be used for storing the wash care instructions, due to the privacy concerns. Hence, additional tags are needed for the wash care instructions, which will increase the cost of the garment. In addition, the consumers need special readers to extract the information stored in the RFID tag. Hence, they will prefer the physical label instead of an electronic label. Furthermore, the use of RFID tags in the garment, may pose health risks to the consumers due to the exposure to radiation.

The other concern related to the use of RFID tag is the electronic waste. An appropriate method should be devised to recycle or reuse the RFID tags, if they are removed at the point of sale. Similarly, if customers remove the chip at some point, proper disposal of the tags is essential, which causes additional worries to the companies. In addition, the other issues related to RFID are the lack of standardisation and high cost. Standardising the technology, design and use of the tags can solve the issues related to lack of standardisation, whereas the technological developments can help to overcome the cost related issues.

Radio frequency identification tags can be used in combination with the registered identification number (RN) system, which is currently in use in some places including the USA. The Federal Trade Commission (FTC) issues the RN, which can be used on the textile label instead of the company name. The use of RN system assists the buyers to easily identify a company from the RN using internet. The RN system uses less space on the label and more space is available for other care instructions. Therefore, the combined use of RFID and RN number systems could further benefit industry and consumers.

The use of RFID tags is still in its infancy stage. Hence, it is too early to conclude the final shape of RFID tag and its impact on consumers. One of the industry observed benefits to the consumers by including washing instructions in the tag. In Germany, RFID chips are being used in nursing homes to help the nurses in correctly sorting the garment according to care instructions. Hence, there is a potential benefit of RFID tags to the consumers. However, all the major drawbacks should be overcome to make RFID tags a success.

3. CUSTOMER RELATIONSHIP MANAGEMENT

It is quite imperative to understand the needs and taste of customers especially in fashion industry to offer them the right product of their match. Moon and Ngai (2008) conducted a study on the use of RFID in fashion retailing and reflected that different customers behaved different at different time in their shopping experience and knowing them in depth regarding their shopping experience was very important so that sales personnel could provide them with personalized services. The interviewee in their study said that if the customer entered into the store and picked few item connected to

RFID technology that would provide signal of their pattern of choices on that day so tailored advice can be provided by sales staff. As competition is intense between the brands so it is important to satisfy the customers by knowing their choice by creating their history of purchases, especially for high-end products. By creating their profile, age group, choice trend, preferred styles etc. the professional suggestion can be provided upon their next purchase so that they feel taken care of. Other significance of using RFID is to be in constant contact with VIP customers, even if the staff member quits the job.

As customers preferences are different so as their buying behavior and pattern, RFID can be used to interact product with the consumer to suit their choices and assist them making purchase decision (Nayak et al. 2015). Subsequently the customers can be offered customized assistance to cater to their choices thus increasing the revenue.

Intelligent clothes racks and smart shelves allowed Kaufhof to gain additional intelligence about the customers' shopping and buying behaviours, which could then be used for further improving product and service offerings. Intelligent clothes racks were equipped with an RFID antenna. Each time a customer or employee removed a fashion item equipped with an RFID transponder from the rack, the movement was automatically entered into the merchandise management system. The moment the article was returned was also registered. The time that lapsed before a customer returned an item to the rack provided valuable information: If a customer put a blouse back on the clothes rack after 5 s, she possibly did not like the material. If 5 min went by, it is likely that the customer tried on the article. *Smart Shelves* offered similar functionalities for stacked goods.

4. CHALLENGES TO RFID

Although RFID technology has already been applied effectively, have certain technological barriers that still need to be overcome to optimize its application. These lacunas are high investment, lack of security and privacy, and some related to the technology of RFID (Wu et al. 2006; Jones et al. 2005; Welbourne et al. 2007).

4.1 High cost

Although there is a great potential of RFID in the local logistics sector, the major drawback is the cost of the RFID tag is higher as compared to barcode system (Ohkubo et al. 2005, Jones et al. 2005). Therefore, industrial leaders are concerned about the return on investment and net profit by investing the extra cost into the existing system. The cost depends on the volume of usage. The lowest cost tags available on the market are as low as 7.2 cents each in volumes of 10 million units or more. The average cost of a RFID tag is around 30–40 cents if the volume is not sufficiently large but the cost is only 4 cents for a barcode. The companies like Wall-mart, Benetton, and Prada are trying to still reduce the price. After the price is reduced all the retailers and manufacturers can implement the technology. Apart from the initial cost, there is cost involved in maintenance and up keeping of the system. But RFID tags will completely replace the barcode system in future.

The use of RFID in supermarkets is worthwhile because the transactions speed in supermarket should be fast and prompt to reduce the queuing time whereas in fashion industry the purchase is not in bulk so transaction speed is not very relevant in fashion industry. So the cost of using RFID tags for economical/low-priced clothes will impact significantly on the final cost of the garment. But where quick scanning is required in case of larger purchases and promptly replenishing merchandise, the use of RFID is very important and trustworthy.

4.2 Security and privacy issues

The security and privacy of the RFID against unauthorized readers is in debate from the very beginning (Sarma et al. 2003; Peris-Lopez et al. 2006; Gao et al. 2004). There is a great challenge to the consumer privacy. The consumers using the product with RFID tags can be traced easily. The RFID tag broadcasts the ID serial number or the EPC to the nearby reader. There is a very high chance of privacy violations. The size of the dress a woman wants can be publicly readable by any nearby scanner. Due to this problem the cloth retailer Benetton has planned to withdraw the use of RFID tags in its apparel products. CASPIAN (Consumers against Supermarket Policy Invasion and Numbering) group, based in US has claimed that RFID tags embedded in items can track the consumers creating unwanted surveillance and an invasion of privacy.

Many security measures have been proposed for RFID in various literatures to overcome the above problem (Peris-Lopez et al. 2006; Osaka et al. 2006). The simplest is the "Kill Tag" approach in which the tag is electronically deactivated after the item is being sold out. The tag also can be removed physically from the product before they are placed on the hand of the consumers. The other approaches are Cryptographic approach, Hash Function approach, Faraday Cage approach, Randomized Hash Lock, backward channel XORing, Active Jamming approach, Regulation approach etc.

4.3 RFID attacks

4.3.1 Tag damage

The RFID tags can be damaged by attackers or due to the improper handling of the material. This attack in turn can prevent the data communication with the reader.

4.3.2 Fake tags

The attackers may replace the original tags with fake ones, which will prevent the data transfer or provide fake information to t the reader.

4.3.3 Security attacks

The attackers can break the security algorithm used in the RFID system or modify the message in the reader.

4.4 Compatibility

A lot of time and money is needed to set up the RFID systems, which may not be economical for many companies. Furthermore, RFID may not be compatible with some existing technologies and even if it can be integrated with existing system the expectancy of 100% data accuracy is still a matter of concern (Smart et al. 2010; Lee and Chan 2009). This becomes more important when the suppliers and buyers have to deal with their overseas counterparts. Some companies are afraid of using the RFID technology as the high investment may be replaced by some other technologies as the technological advancements in near future may bring better technologies than RFID.

4.5 Technology

As the RFID is based on the concept of Radio Frequency, it can be interfered with other radio transmissions, metals, liquids etc. (Bunduchi et al. 2011; Wu et al. 2006; Ni et al.2011). The degree of interference depends upon the frequency of the tag and the usage environment, which may lead to inaccurate reading. For example the use of RFID readers to read the price tags of purchased merchandise at the checkout counters may be interfered by radio waves from other merchandise. In addition, the merchandise are made in different countries and may be packed with different material.

4.6 Lack of standardization

Radio frequency identification is at the infancy stage and there are many hurdles ahead of it. There are many versions of RFID that operate at different frequencies and need different software and readers. So the need is to be agreed upon one or group of frequencies to have interoperability between the manufacturers, retailers and distributors. Currently, two organisations: EPC global and International Standards Organization (ISO) are working to standardize the use of RFID in UHF spectrum (Wu et al. 2006). Although, standards such as EPC class 1 G2 protocol and ISO 18000-6 are being released, both the standards are yet not finalised and lack compatibility with each other.

5. RESEARCH ON RFID

Various research have been done on the technology including tags, antennae and readers of RFID (Nikitin et al. 2005; Ukkonen et al. 2012; Kellomäki et al. 2012; Occhiuzzi et al.2010; Gimpel et al. 2004); communications infrastructure (Srivastava 2010; Zuo 2010; Welbourne et al. 2007); logistics (Wang and Wang 2009; Kapoor et al. 2009); security (Fishkin et al. 2005; Simson et al. 2005; Thiesse 2007; Yeo and Kim 2005; Knospe and Pohl 2004); and standardization (Moon and Ngai 2008; Bolic et al. 2010).

Although several challenges are present for the adoption of RFID, they can be overcome with the technical developments. Similar to other emerging technologies, RFID will become a reality in fashion and textiles over time. Research and development can help in mitigating the associated problems with the use of RFID. The technological issues such as 100% reliability can be achieved by: (1) installing multiple antennas with different orientations per reader, (2) customizing the tag and antenna design for each material type and (3) installing multiple readers to increase readability of the tags. Reading multiple tags simultaneously is possible by the use of binary search technique. Repetition of reading the tags can be avoided by putting the tags already read into sleep mode. Research and development will help in maximizing the antenna power pattern and the chances of successful reading (Weber 2010).

Research is needed to explore how RFID technology can be used to improve the inventory management performance by reducing forecast error and inventory discrepancy. Various models need to be developed to establish a suitable benchmark for calculating the value of the RFID system to improve the inventory management (Burmester and De Medeiros2007; Nayak and Padhye 2011). Research is needed to manage inventory operations in a supply chain where both the systems (bar coding and RFID) are working.

Furthermore, for successful application of RFID, existing research should focus on a host of disciplines, such as electronic engineering, information systems, computer science, and business strategy (Bhattacharya et al. 2007). Prior to the research and development, it is vital to determine the principal concerns of RFID, whether technological, application related, or security related. Maintaining security of the RFID tags is a challenging job. The improvement in information technology and design of smart RFID tags can help to overcome this challenge.

6. CONCLUSION AND FUTURE TRENDS OF RFID IN APPAREL INDUSTRY

The use of RFID systems in textiles is rapidly increasing. In future almost all the manufacturers, distributors and retailers in textile business will be using the RFID system. Although, RFID systems cannot completely replace the barcode technology, due to higher cost but the accuracy, speed and the return on investment is high in RFID system. RFID system is a promising technology, which will become ubiquitous in the future helping organizations to solve problems in supply chain management, security, personal identification and object tracking.

The retailers, manufacturers and consumer goods companies like CVS, Tesco, Prada, Benetten, Wal-mart and Procter & Gamble now implementing the technology and exploring the impact of the technology on their business. Others can take the advantages of RFID. The basic of success lies in understanding the technology and other features to minimize the potential problems. It is high-time that the apparel industries should start utilizing the new RFID technology in various applications such as manufacturing, storing purposes.

There is a great potential of RFID in various applications relater to clothing and textiles, manufacturing, retailing and inventory management. The direct benefits have been shown by the companies such as GAP, the US apparel company, who increased their sales by 2% by using RFID to reduce stock-outs. Another well-known US apparel company, J Crew, increased the speed of inventory keeping time five to eight times by the use of RFID.

The apparel industry is characterized by short product life cycle, frequent style changes, a wide range of product designs and hence input materials; variable production volumes; high competitiveness and often high demand on product quality. In addition to these challenges, the use of RFID increases the cost and maintaining privacy creates extra challenges. The organizations dealing with textile and apparel products should start with the tracking of expensive goods as the reduction in the labor and theft of these items can compensate the cost of these items. RFID has advantages in speed, accuracy and convenient over traditional barcode systems. Therefore, in the future, RFID tagging will gradually replace traditional barcode systems and facilitate supply chain and inventory management.

Important technical challenges still need to be resolved for fully successful operation of RFID in textile and apparel manufacturing. It will take some years may be a decade or so, before the full benefits of RFID technology is realized. We will increasingly see how RFID technology can extend the ability of computers in combination with internet to sense and respond to the physical world. In future, RFID will make computing an unobtrusive, intuitive part of everyday life, as refreshing as a walk through the nature.

The largest challenge to the widespread use of RFID is the cost in textile and apparel manufacturing. These challenges can be reduced by increasing the volume of usage. In addition standardization and technological developments can help in reducing the cost. RFID chips can be developed, which can work in different places for different applications. These ideal RFID chips will require a frequency independent capability and would be integrated with read/write memory. In addition, the software needs to be customized to accommodate the number of different uses. Recently, all the merchandise uses the barcodes for identification and at the point of sale, which are easily printed and attached to the products within a very short period. However, finalization of the design and preparation of RFID tags is time consuming process.

The issues related to standardization can be resolved by the effort of governing bodies to develop various standards related to their use, manufacture and technical specification. Technological developments will also assist in this direction to achieve wide adoptability of RFID tags. In addition, throughout a supply chain RFID-friendly infrastructure development is a time consuming process.

From the above discussions, it is clear that RFID technology has a very bright future in the textile and clothing industry. New areas of applications will emerge by the technical advancements and we will see more RFID applications in departments where they are currently not being used. This paper will be a good resource for those interested in RFID research, including academic researchers and industry practitioners, and will provide guideline to stimulate further interest in this area.

AUTHORS' CONTRIBUTIONS

RN designed the paper, performed the literature survey and completed the paper. AS completed the application of RFID in textile and fashion retailing section and helped in feedback and comments. RP and LW guided during the paper design and helped in feedback and comments in final format of the paper. All authors read and approved the final manuscript.

COMPETING INTERESTS

The authors declare that they have no competing interests.

REFERENCES

1. Age, C. S. (2005). Research confirms Wal-Mart's RFID benefit. *Chain Store Age, 81*, 80.

2. Ayoade, J. (2007). Roadmap to solving security and privacy concerns in RFID systems.*Computer Law & Security Review, 23*, 555–561.CrossRef

3. Bagchi, U., Guiffrida, A., O'Neill, L., Zeng, A., & Hayya, J. (2007). The effect of RFID on inventory management and control. *Teends in supply chain design and management*. New York: Springer.

4. Barnett, R. E., & Liu, J. (2008). An EEPROM programming controller for passive UHF RFID transponders with gated clock regulation loop and current surge control. *IEEE Journal of Solid-State Circuits, 43*, 1808–1815.CrossRef

5. Bhattacharya, M., Chu, C. H., & Mullen, T. (2007). RFID implementation in retail industry: Current status, issues, and challenges. In *Proceedings of 38th Annual Meeting of the Decision Sciences Institute, Phoenix, AZ, 2007* (pp. 2171–2176). Citeseer.

6. Bogdanov, A., Leander, G., Paar, C., Poschmann, A., Robshaw, M. J., & Seurin, Y. (2008).*Hash functions and RFID tags: Mind the gap. Cryptographic hardware and embedded systems (CHES 2008)*. New York: Springer.

7. Bolic, M., Simplot-Ryl, D., & Stojmenovic, I. (2010). *RFID systems: Research trends and challenges*. Hoboken: Wiley.

8. Bottani, E., Ferretti, G., Montanari, R., & Rizzi, A. (2009). The impact of RFID technology on logistics processes of the fashion industry supply chain. *International Journal of RF Technologies: Research and Applications, 1*, 225–252.CrossRef

9. Bouet, M., & Dos Santos, A. L. (2008). RFID tags: Positioning principles and localization techniques. In *Wireless Days, 2008 (WD'08), 1st IFIP, 2008* (pp. 1–5). IEEE.

10. Bunduchi, R., Weisshaar, C., & Smart, A. U. (2011). Mapping the benefits and costs associated with process innovation: The case of RFID adoption. *Technovation, 31*, 505–521.CrossRef

11. Burmester, M., & de Medeiros, B. (2007). RFID security: Attacks, countermeasures and challenges. In *The 5th RFID Academic Convocation, The RFID Journal Conference*.http://www. cs.fsu.edu/~burmeste/133.pdf.

12. Chao, C. C., Yang, J. M., & Jen, W. Y. (2007). Determining technology trends and forecasts of RFID by a historical review and bibliometric analysis from 1991 to 2005. *Technovation,27*, 268–279.

13. Chen, J. V., & Pfleuger Jr, P. (2008). RFID in retail: A framework for examining consumers' ethical perceptions. *International Journal of Mobile Communications, 6*, 53–66.

14. Christopher, M., Lowson, R., & Peck, H. (2004). Creating agile supply chains in the fashion industry. *International Journal of Retail & Distribution Management, 32*, 367–376.

15. Costin, A., Pradhananga, N., & Teizer, J. (2012). Leveraging passive RFID technology for construction resource field mobility and status monitoring in a high-rise renovation project. *Automation in Construction, 24*, 1–15.

16. de Kok, A., van Donselaar, K., & van Woensel, T. (2008). A break-even analysis of RFID technology for inventory sensitive to shrinkage. *International Journal of Production Economics, 112*, 521–531.

17. Delen, D., Hardgrave, B. C., & Sharda, R. (2007). RFID for better supply-chain management through enhanced information visibility. *Production and Operations Management, 16*, 613–624.

18. Dimitriou, T. (2005). A lightweight RFID protocol to protect against traceability and cloning attacks. In *First international conference on security and privacy for emerging areas in communications networks, 2005 (SecureComm 2005)* (pp. 59–66). IEEE.

19. dos Santos, B. L., & Smith, L. S. (2008). RFID in the supply chain: Panacea or Pandora's box? *Communications of the ACM, 51*, 127–131.

20. Dutta, A., Lee, H. L., & Whang, S. (2007). RFID and operations management: Technology, value, and incentives. *Production and Operations Management, 16*, 646–655.

21. Fishkin, K. P., Roy, S., & Jiang, B. (2005). *Some methods for privacy in RFID communication. Security in ad-hoc and sensor networks*. New York: Springer.

22. Gao, X., Xiang, Z., Wang, H., Shen, J., Huang, J., & Song, S. (2004). An approach to security and privacy of RFID system for supply chain. In *IEEE international conference on e-commerce technology for dynamic e-business, 2004* (pp. 164–168). IEEE.

23. García, A., Chang, Y., Abarca, A., & Oh, C. (2007). RFID enhanced MAS for warehouse management. *International Journal of Logistics, 10*, 97–107.

24. Gimpel, S., Mohring, U., Muller, H., Neudeck, A., & Scheibner, W. (2004). Textile-based electronic substrate technology. *Journal of Industrial Textiles, 33*, 179–189.

25. Goodrum, P. M., McLaren, M. A., & Durfee, A. (2006). The application of active radio frequency identification technology for tool tracking on construction job sites. *Automation in Construction, 15*, 292–302.

26. Hardgrave, B. C., Aloysius, J., & Goyal, S. (2009). Does RFID improve inventory accuracy? A preliminary analysis. *International Journal of RF Technologies: Research and Applications, 1*, 44–56.

27. Jaselskis, E. J., Anderson, M. R., Jahren, C. T., Rodriguez, Y., & Njos, S. (1995). Radio frequency identification applications in construction industry. *Journal of Construction Engineering and Management, 121*, 189–196.

28. Jihui, G., Fuhai, L., Bing, D., Peng, W., & Lehong, H. (2011). A design of RFID tag antenna for clothing. In *Cross Strait Quad-Regional Radio Science and Wireless Technology Conference (CSQRWC), 2011* (pp. 1075–1077). IEEE.

29. Jin, G. Y., Lu, X. Y., & PARK, M. S. (2006). An indoor localization mechanism using active RFID tag. In *IEEE International conference on sensor networks, ubiquitous, and trustworthy computing, 2006* (p. 4). IEEE.

30. Jones, P., Clarke-Hill, C., Hillier, D., & Comfort, D. (2005). The benefits, challenges and impacts of radio frequency identification technology (RFID) for retailers in the UK.*Marketing Intelligence & Planning, 23*, 395–402.CrossRef

31. Kapoor, G., Zhou, W., & Piramuthu, S. (2009). Challenges associated with RFID tag implementations in supply chains. *European Journal of Information Systems, 18*, 526–533.

32. Kärkkäinen, M. (2003). Increasing efficiency in the supply chain for short shelf life goods using RFID tagging. *International Journal of Retail & Distribution Management, 31*, 529–536.

33. Kellomäki, T., Virkki, J., Merilampi, S., & Ukkonen, L. (2012). Towards washable wearable antennas: a comparison of coating materials for screen-printed textile-based UHF RFID tags. *International Journal of Antennas and Propagation, 2012.*

34. Knospe, H., & Pohl, H. (2004). RFID security. *Information Security Technical Report, 9*, 39–50.

35. Landt, J. (2005). The history of RFID. *Potentials, IEEE, 24*, 8–11.

36. Lee, C., & Chan, T. (2009). Development of RFID-based reverse logistics system. *Expert Systems with Applications, 36*, 9299–9307.

37. Lee, C., Choy, K. L., Ho, G. T., & Law, K. (2013). A RFID-based resource allocation system for garment manufacturing. *Expert Systems with Applications, 40*, 784–799.

38. Legnani, E., Cavalieri, S., Pinto, R., & Dotti, S. (2011). The potential of RFID technology in the textile and clothing industry: Opportunities, requirements and challenges. In *Unique radio innovation for the 21st century*. New York: Springer.

39. Li, X., Zhang, Y., & Amin, M. G. (2009). Multifrequency-based range estimation of RFID tags. In *IEEE international conference on RFID, 2009* (pp. 147–154). IEEE.

40. Liu, J., Gao, W., Wang, H., Jiang, H., & Li, Z. (2010). Development of bobbin tracing system based on RFID technology. *The Journal of the Textile Institute, 101*, 925–930.

41. Liu, F., & Miao, Z. (2006). The application of RFID technology in production control in the discrete manufacturing industry. In *IEEE international conference on video and signal based surveillance, 2006 (AVSS'06)* (p. 68). IEEE.

42. Loebbecke, C., & Huyskens, C. (2008). A competitive perspective on standard-making: Kaufhof's rfid project in fashion retailing. *Electronic Markets, 18*, 30–38.

43. Michael, K., & Mccathie, L. (2005). The pros and cons of RFID in supply chain management. In *International conference on mobile business, 2005 (ICMB 2005)* (pp. 623–629). IEEE.

44. Miles, S. B., Sarma, S. E., & Williams, J. R. (2008). *RFID technology and applications*. Cambridge: Cambridge University Press.

45. Moon, K., & Ngai, E. (2008). The adoption of RFID in fashion retailing: A business value-added framework. *Industrial Management & Data Systems, 108*, 596–612.

46. Myny, K., Steudel, S., Smout, S., Vicca, P., Furthner, F., van der Putten, B., et al. (2010). Organic RFID transponder chip with data rate compatible with electronic product coding.*Organic Electronics, 11*, 1176–1179.

47. Nath, B., Reynolds, F., & Want, R. (2006). RFID technology and applications. *IEEE Pervasive Computing, 5*, 22–24.

48. Nayak, R., Chatterjee, K., Khurana, G., & Khandual, A. (2007). RFID: Tagging the new era.*Man-Made Textiles in India, 50*, 174–177.

49. Nayak, R., & Padhye, R. (2011). Application of modelling and simulation in smart and technical textiles. In A. Patanaik (Ed.), *Modeling and simulation in fibrous materials: techniques and applications.* Huntington: Nova Science.

50. Nayak, R., & Padhye, R. (2014). The care of apparel products. In R. Sinclair (Ed.), *Textiles and fashion: materials, design and technology.* Kidlington: Elsevier.

51. Nayak, R., & Padhye, R. (2015a). Introduction: The apparel industry. In R. Nayak & R. Padhye (Eds.), *Garment manufacturing technology.* Cambridge: Elsevier.

52. Nayak, R., & Padhye, R. (2015b). Care labeling of clothing. In R. Nayak & R. Padhye (Eds.),*Garment manufacturing technology.* Amsterdam: Elsevier.

53. Nayak, R., Padhye, R., Wang, L., Chatterjee, K. & Gupta, S. (2015). The role of mass customisation in the apparel industry. *International Journal of Fashion Design, Technology and Education,* 1–11.

54. Ni, L. M., Liu, Y., Lau, Y. C., & Patil, A. P. (2004). LANDMARC: Indoor location sensing using active RFID. *Wireless Networks, 10*, 701–710.

55. Ni, L. M., Zhang, D., & Souryal, M. R. (2011). RFID-based localization and tracking technologies. *Wireless Communications, IEEE, 18*, 45–51.

56. Nikitin, P.V., Lam, S., & Rao, K. (2005). Low cost silver ink RFID tag antennas. In*Antennas and propagation society international symposium, 2005* (pp. 353–356) IEEE.

57. Nikitin, P.V., Parks, A., & Smith, J.R. (2013). RFID-Vox: A tribute to Leon Theremin. In*Wirelessly powered sensor networks and computational RFID*. New York: Springer.

58. Occhiuzzi, C., Cippitelli, S., & Marrocco, G. (2010). Modeling, design and experimentation of wearable RFID sensor tag. *IEEE Transactions on Antennas and Propagation, 58*, 2490–2498.

59. Ohkubo, M., Suzuki, K., & Kinoshita, S. (2005). RFID privacy issues and technical challenges. *Communications of the ACM, 48*, 66–71.CrossRef

60. Osaka, K., Takagi, T., Yamazaki, K., & Takahashi, O. (2006). An efficient and secure RFID security method with ownership transfer. In *International conference on computational intelligence and security, 2006* (pp. 1090–1095). IEEE.

61. Peris-Lopez, P., Hernandez-Castro, J. C., Estevez-Tapiador, J. M., & Ribagorda, A. (2006). RFID systems: A survey on security threats and proposed solutions. In *Personal wireless communications, 2006* (pp. 159–170). New York: Springer.

62. Phillips, T., Karygiannis, T., & Kuhn, R. (2005). Security standards for the RFID market.*Security & Privacy, IEEE, 3*, 85–89.

63. Piramuthu, S., Wochner, S., & Grunow, M. (2014). Should retail stores also RFID-tag 'cheap'items? *European Journal of Operational Research, 233*, 281–291.

64. Rao, K. S., Nikitin, P. V., & Lam, S. F. (2005). Antenna design for UHF RFID tags: A review and a practical application. *IEEE Transactions on Antennas and Propagation, 53*, 3870–3876.

65. Rekik, Y., Sahin, E., & Dallery, Y. (2008). Analysis of the impact of the RFID technology on reducing product misplacement errors at retail stores. *International Journal of Production Economics, 112*, 264–278.

66. Reynolds, K. E., Folse, J. A. G., & Jones, M. A. (2006). Search regret: Antecedents and consequences. *Journal of Retailing, 82*, 339–348.

67. Roberti, M. (2006). RFID is fit to track clothes. *Chain Store Age, 82*, 158.

68. Roberts, C. M. (2006). Radio frequency identification (RFID). *Computers & Security, 25*, 18–26.

69. Roussos, G. (2006). Enabling RFID in retail. *Computer, 39*, 25–30.

70. Sarma, S., Brock, D., & Engels, D. (2001). Radio frequency identification and the electronic product code. *IEEE Micro, 21*, 50–54.

71. Sarma, S. E., Weis, S. A., & Engels, D. W. (2003). RFID systems and security and privacy implications. In *Cryptographic hardware and embedded systems (CHES 2002).* New York: Springer.

72. Simson, L., Juels, A., & Pappu, R. (2005). RFID privacy: An overview of problems and proposed solutions. In *IEEE symposium on security & privacy, 2005* (pp. 34–43).

73. Smart, A. U., Bunduchi, R., & Gerst, M. (2010). The costs of adoption of RFID technologies in supply networks. *International Journal of Operations & Production Management, 30*, 423–447.

74. Srivastava, B. (2010). Critical management issues for implementing RFID in supply chain management. *International Journal of Manufacturing Technology and Management, 21*, 289–307.

75. Sun, C. (2012). Application of RFID technology for logistics on internet of things. *AASRI Procedia, 1*, 106–111.

76. Tan, H. (2008). The application of RFID technology in the warehouse management information system. In *International symposium on electronic commerce and security, 2008* (pp. 1063–1067). IEEE.

77. Tedjini, S., Perret, E., Vena, A., & Kaddour, D. (2012). Mastering the electromagnetic signature of chipless RFID tags. In N. Chandra Karmakar (Ed.), *Chipless and conventional radio frequency identification: Systems for ubiquitous tagging* (pp. 146–174). IGI Global, USA

78. Tesoriero, R., Gallud, J., Lozano, M., & Penichet, V. (2008). Using active and passive RFID technology to support indoor location-aware systems. *IEEE Transactions on Consumer Electronics, 54*, 578–583.

79. Thiesse, F. (2007). RFID, privacy and the perception of risk: A strategic framework. *The Journal of Strategic Information Systems, 16*, 214–232.

80. Ukkonen, L., Sydanheimo, L., & Rahmat-Samii, Y. (2012). Sewed textile RFID tag and sensor antennas for on-body use. In *6th European conference on antennas and propagation (EUCAP), 2012* (pp. 3450–3454). IEEE.

81. Vowels, S. A. (2006). A straegic case for RFID: An examination of Wal-mart and its supply chain.

82. Wang, Z. G., Tang, R. Z., Sheng, W. L., & Wang, G. D. (2006). Research on RFID-based production logistics management techniques

with application in garment industry. In*International technology and innovation conference, 2006 (ITIC 2006)* (pp. 383–387). IET.

83. Wang, L., & Wang, G. (2009). RFID-driven global supply chain and management.*International Journal of Computer Applications in Technology, 35*, 42–49.

84. Want, R. (2006). An introduction to RFID technology. *Pervasive Computing, IEEE, 5*, 25–33.

85. Weber, R. H. (2010). Internet of things-new security and privacy challenges. *Computer Law & Security Review, 26*, 23–30.

86. Weinstein, R. (2005). RFID: A technical overview and its application to the enterprise. *IT professional, 7*, 27–33.

87. Welbourne, E., Balazinska, M., Borriello, G., & Brunette, W. (2007). Challenges for pervasive RFID-based infrastructures. In *Fifth annual IEEE international conference on pervasive computing and communications workshops, 2007 (PerCom Workshops' 07)*(pp. 388–394). IEEE.

88. Whitaker, J., Mithas, S., & Krishnan, M. S. (2007). A field study of RFID deployment and return expectations. *Production and Operations Management, 16*, 599–612.

89. White, G., Gardiner, G., Prabhakar, G. P., & ABD RAZAK, A. (2007). A comparison of barcoding and RFID technologies in practice. *Journal of Information, Information Technology and Organizations, 2*, 119–132.

90. Wu, D. L., Ng, W. W., Yeung, D. S., & Ding, H. L. (2009). A brief survey on current RFID applications. In *International conference on machine learning and cybernetics, 2009* (pp. 2330–2335). IEEE.

91. Wu, N. C., Nystrom, M., Lin, T. R., & Yu, H. C. (2006). Challenges to global RFID adoption. *Technovation, 26*, 1317–1323.

92. Yeo, S. S., & Kim, S. K. (2005). Scalable and flexible privacy protection scheme for RFID systems. In *Security and privacy in ad hoc and sensor networks*. New York: Springer.

93. Zhang, Y., & Amin, M. G. (2006) Localization and tracking of passive RFID tags. In*Defense and security symposium, 2006* (pp. 624809–624811). International Society for Optics and Photonics.

94. Zuo, Y. (2010). Survivable RFID systems: Issues, challenges, and techniques. *IEEE Transactions on Systems, Man, and Cybernetics, Part C: Applications and Reviews, 40*, 406–418.

CHAPTER 9

Influence of Immersion Conditions on The Tensile Strength of Recycled Kevlar®/Polyester/Low-Melting-Point Polyester Nonwoven Geotextiles through Applying Statistical Analyses

Jing-Chzi Hsieh [1], Jia-Hsun Li [2], Ching-Wen Lou [3], Chien-Teng Hsieh [4], Wen-Hao Hsing [5], Yi-Jun Pan [6] and Jia-Horng Lin [7,8,9,*]

[1]Department of Land Management, Feng Chia University, No. 100, Wenhwa Rd., Seatwen, Taichung 40724, Taiwan

[2]Department of Ph. D Program in Civil and Hydraulic Engineering, Feng Chia University, No. 100, Wenhwa Rd., Seatwen, Taichung 40724, Taiwan

[3]Institute of Biomedical Engineering and Materials Science, Central Taiwan University of Science and Technology, No. 666, Buzih Road, Beitun District, Taichung 40601, Taiwan

[4]Department of Fashion Design and Merchandising, Shih Chien University Kaohsiung Campus, Kaohsiung 845, Taiwan

[5]Department of Textile Engineering, Chinese Culture University, Taipei 11114, Taiwan

[6]Department of Material and Fiber, Oriental Institute of Technology, New Taipei 22061, Taiwan

[7]Laboratory of Fiber Application and Manufacturing, Department of Fiber and Composite Materials, Feng Chia University, No. 100, Wenhwa Rd., Seatwen, Taichung 40724, Taiwan

[8]School of Chinese Medicine, China Medical University, No. 91, Xueshi Rd., North Dist., Taichung 40402, Taiwan

[9] Department of Fashion Design, Asia University, No. 500, Liufeng Rd., Wufeng Dist., Taichung 41354, Taiwan

ABSTRACT

The recycled Kevlar®/polyester/low-melting-point polyester (recycled Kevlar®/PET/LPET) nonwoven geotextiles are immersed in neutral, strong acid, and strong alkali solutions, respectively, at different temperatures for four months. Their tensile strength is then tested according to various immersion periods at various temperatures, in order to determine their durability to chemicals. For the purpose of analyzing the possible factors that influence mechanical properties of geotextiles under diverse environmental conditions, the experimental results and statistical analyses are incorporated in this study. Therefore, influences of the content of recycled Kevlar® fibers, implementation of thermal treatment, and immersion periods on the tensile strength of recycled Kevlar®/PET/LPET nonwoven geotextiles are examined, after which their influential levels are statistically determined by performing multiple regression analyses. According to the results, the tensile strength of nonwoven geotextiles can be enhanced by adding recycled Kevlar® fibers and thermal treatment.

KEYWORDS

recycled Kevlar fibers; low-melting-point polyester; nonwoven geotextiles; tensile strength; statistical analysis

1. INTRODUCTION

Nonwoven geotextiles are commonly used in geotechnical, hydraulic engineering, and environmentally protective fields, and have functions of filtration, drainage, separation, protection, and reinforcement [1,2,3,4]. Thermoplastic materials are made into geotextiles via mechanical or chemical bonding methods. The used synthetic fibers include polypropylene (PP), polyester (PET), polyethylene, polyvinyl chloride, and polyamide, and PP and PET are the most commonly used. In comparison to PP, PET has greater mechanical properties, lower creep property and better thermal resistance. PET has a melting point of 259–269 °C [5,6].

Furthermore, Thwe *et al.* [7] and Jawaid *et al.* [8] reported that, with appropriate material designs, uses of two or more fibers to form composites can compensate for the drawbacks of using a single material, and further provide the products with greater mechanical properties, physical properties, and a production that consumes low energy. In recent years, there have been a great deal of studies examining the ability of high modulus fibers to fortify nonwoven composites, while glass-fiber woven fabrics, aramide fibers, and carbon fibers are currently used to reinforce composites. In the study by Li *et al.*, Kevlar®/high-strength polyamide 6/low-melting PET nonwoven and low-melting PET/PET nonwoven served as the top and bottom layers, enclosing a layer of glass woven fabric to form three-dimensional compound fabrics. The results indicate that increasing polyamide fibers benefits the static and dynamic punctures of the compound fabrics [9]. Yan *et al.* also published a study that combined a high-density flexible polyurethane foam as the interlayer with a recycled Kevlar®/low-melting-point polyester nonwoven fabric as the top layer and a glass woven fabric as the bottom layer. The tensile strength and tear resistance of the resulting sandwiches were distinctly increased [10].

Therefore, this study combines recycled Kevlar® fibers and low-melting-point fibers to form geotextiles in order to improve the mechanical and physical properties of PET geotextiles. Kevlar® fibers are featured by having a high modulus, high tenacity, chemical resistance, impact resistance, thermal resistance, and abrasion resistance. As a result, Kevlar® fibers have commonly been used for body armor, stab-resistant protective products, tires, steamships, light aircraft, and submarine cables. Meanwhile, there is also by-product of Kevlar® selvages, which are mostly burned [11]. The burning process results in greater energy consumption and worst-case forms of pollutions. It benefits the environment and new products if Kevlar®selvages can be recycled for other applications. Li *et al.* indicated that using recycled Kevlar® fibers resulted in cost reduction, environmental benefits, and mechanical property reinforcement [12]. In addition, the incorporation of low-melting-point PET (LPET) fibers improves the properties of nonwoven products and the pollution problems that are caused by massive usage of adhesives. LPET fibers are modified composite fibers and have a sheath-core structure. Their outer layer has a melting point of 110 °C, and LPET fibers are thus commonly used for thermal consolidation. Thermal treatment melts part of LPET fibers, thereby yielding the function of adhesives. Relative literature sources also confirm that the incorporation of LPET fiber reinforces the mechanical properties of the nonwoven fabrics

[13,14]. Therefore, this study combines PET fibers, Kevlar® fibers, and LPET fibers to form nonwoven geotextiles. This product design for geotextiles emphasizes both their initial performance and their durability. Geotextiles are commonly closely associated with aqueous acid/alkali solutions or dissolved oxygen. It is thus essential to evaluate the chemical resistance of geotextiles that are greatly dependent on their polymer compositions, manufacturing parameters, fabric structures, as well as external factors, including liquid composition and *in situ* conditions [15,16,17,18].

In the study by East *et al.*, polyester and polypropylane geotextiles were immersed in saline, strong acidic solution (pH 3), and strong alkaline solution (pH 10) at room temperature and 95 °C, thereby examining the influence of the accelerated aging experiment on the properties of geotextiles. The test results indicated that the influence of groundwater chemistry on the durability of geotextiles was important when geotextiles were used in hazardous waste landfills, slopes, subgrades and earth retention systems [16]. As indicated in ISO 13434 (Geosynthetics—Guidelines for the Assessment of Durability), the pH value of the environment reaches 9–11 for specified soils or a special construction process [19]. In addition, the pH value of soils in Taiwan is between 4 and 6. Due to a limited time, this study thus uses extreme conditions to examine the influence of adding recycled Kevlar® fibers on tensile strength of nonwoven geotextiles. In order to examine possible factors that change the durability and mechanical properties of geotextiles in diverse environments, this study immerses recycled Kevlar®/PET/LPET nonwoven geotextiles in neutral, acid, and alkaline solutions that is at different temperatures, and measures their tensile strength every month for four consecutive months. Strong alkali and acid solutions are thus used for a maximum of a 4 month immersion in order to exert as much severe influence over the tensile strength of the nonwoven geotextiles. Statistical analyses are then used to analyze the differences in the tensile strength of the nonwoven fabrics, in relation to fiber ratios, thermal treatment, and immersion periods. The influential levels of different variables that are demonstrated on recycled Kevlar®/PET/LPET nonwoven geotextiles are determined by using multiple regression analyses. The experimental results can serve as a reference for designing novel geotextile products with their associated environmental requirements.

2. EXPERIMENTAL SECTION

2.1. Materials

Recycled Kevlar® fibers are obtained from Kevlar® uni-directional selvages (Dupont, Pierce County, WA, USA) that include 2820 Denier (D) K129,

1000 D K29, and 2160 D K49 multi-filaments. The length of the Kevlar® fiber ranges from 50 mm to 60 mm. The three-dimensional, crimped, hollow polyester (PET) fiber (Far Eastern New Century, Taipei, Taiwan) has a fineness of 7 D and a length of 64 mm. Low-melting-point PET (LPET) fiber (Far Eastern New Century, Taipei, Taiwan) has a fineness of 4 D and a length of 51 mm. LPET has a sheath-core structure. The sheath is PET fiber with a melting-point of 110 °C and the core is PET fiber with a melting-point of 260 °C.

2.2. Sample Preparation

Kevlar® selvages are processed with opening in order to obtain Kevlar® fibers. Kevlar® fibers are high strength and the machine for nonwoven fabrics also has limits in capacity of processing materials. Kevlar® fibers are thus added with 5 weight percent (wt. %) in this trial experiment, thereby obtaining the optimal production. Kevlar® fibers, PET fibers, and LPET fibers are mixed with various ratios of 0/80/20, 5/75/20, 10/70/20, 15/65/20, and 20/60/20. They are processed with opening, blending, carding, lapping, and needle-punching, in order to form recycled Kevlar®/PET/LPET nonwoven geotextiles with a basis weight of 200 ± 10 g/m^2. The geotextiles are then thermally treated through a calendar at 120 °C. This process melts the sheath structure of LPET fibers and causes thermal bonding points.

2.3. Testing Methods

2.3.1. Tensile Strength of Nonwoven Geotextiles under Diverse Environmental Conditions

Jeon *et al.* examined the tensile strength retention of high-tenacity PET yarn textile geogrids as specified in ASTM D5322, in order to evaluate their chemical resistance [17]. Lodi *et al.* used the same measurement to examine the mechanical and thermal properties of the chemically degraded polyvinyl chloride and high density polyethylene geo-membranes [20]. Therefore, this study proposes nonwoven geotextiles and evaluates their tensile strength and variations in weight. As statistical analyses need continuous variables, the yielded tensile strength data is used for comparison, not tensile strength retention. According to ASTM D5322–98 (Standard Practice for Laboratory Immersion Procedures for Evaluating the Chemical Resistance of Geosynthetics to Liquids) [21], samples are immersed in solutions at different pH and temperature conditions for four months. The solutions for this test are acetic acid and sodium hydroxide as specified in ASTM D543–14 (Standard Practices for Evaluating The Resistance of Plastics to Chemical Reagents) [22]. The solution for the control group is deionized water. Table 1 shows the specifications of three solutions as well as the test conditions. Recycled Kevlar®/PET/LPET nonwoven geotextiles are cut along the cross machine direction (CD) and along the machine direction (MD). Samples have a size of 180 mm × 25.4 mm. Samples are weighed before they are immersed in the solutions for different periods (1, 2, 3, and 4 months). After the immersion for various periods, samples are removed from the solutions, rinsed with deionized water, thermally dried at 30 °C for 24 h, and weighed again. They are respectively tested for tensile strength by using an Instron 5566 universal tester (Instron, Norfolk County, MA, USA), as specified in ASTM D5035-11 [23]. The tensile test speed is 305 ± 13 mm/min, and the distance between clamps is 75 mm. The tensile strength along the CD and MD for different samples is tested with their maximum breaking force being recorded as the tensile strength.

Table 1. The immersion conditions on the tensile strength of recycled Kevlar®/polyester/low-melting-point polyester (recycled Kevlar®/PET/LPET) nonwoven geotextile.

Solutions	Chemical Formula	Solution Concentration	pH	Temperature (°C)	Immersion Period (Months)
Acetic acid	CH_3COOH	5%	3		
Sodium hydroxide	NaOH	10%	13	23, 50	1, 2, 3, 4
Deionized water	H_2O	-	6		

2.3.2. Fourier-Transform Infrared Spectroscopy (FT-IR) Analysis

FT-IR spectra are recorded by using a Lab Solutions IR software program (Version 1.2, Shimadzu Corporation, Kyoto, Japan, 2000) on an IRAffinity-1S FT-IR spectrometer (Shimadzu Corporation, Kyoto, Japan), using the Attenuated Total Reflection (ATR) mode, in the range of 4000–600 cm^{-1} with a resolution of 8 cm^{-1}. Each spectrum is obtained from 20 successive scans.

2.4. Statistical Analyses

Test result data is analyzed with the incorporation of a *t*-test, one-way ANOVA, and multiple regression analyses of a SPSS statistics software program (Version 20.0, International Business Machines Corporation,

Armonk, NY, USA, 2011). Both*t*-test and one-way ANOVA are for the analyses of variance. The α (alpha level) is commonly set as 5% with a confidence interval of 95%. When $p < 0.05$ is sustained, the results indicate significance. For specific analyses of variations between groups that is exemplified by the statistically significant differences in one-way ANOVA, this study then uses the post hoc Scheffe's test in order to examine between what specified parameters that the statistically significant difference is present. Finally, multiple regression analyses are used to analyze the influences of multiple independent variables on a specified dependent variable. When the significance is below 0.05, the results stand for a statistical significance that enables β-distribution to determine the influential level of independent variables on the dependent variables.

3. RESULTS AND DISCUSSION

3.1. Effects of Fiber Ratios on Tensile Strength

Table 2 shows the contents of Kevlar® fibers of recycled Kevlar®/PET/LPET nonwoven geotextiles. The geotextiles without immersion in any solutions have a tensile strength that is proportional to the content of recycled Kevlar® fibers. In comparison to the control group containing zero Kevlar® fibers, the experimental groups containing 15 wt. %–20 wt. % of Kevlar® fibers exhibit statistically significant differences at (F (4, 20) = 5.201, $p < 0.05$). In addition, PET fibers have a lower alkali resistance, and therefore, one-way ANOVA and post hoc Scheffe's tests are performed on the recycled Kevlar®/PET/LPET nonwoven geotextiles in order to determine the influence of Kevlar® contents on their tensile strength. According to one-way ANOVA, the differences are statistically significant ($p < 0.05$) among different contents of recycled Kevlar® fibers, indicating that the contents of Kevlar® pertain to tensile strength. Next, Scheffe's test is used to analyze the differences in tensile strength among nonwoven geotextiles containing various contents of Kevlar® fibers that are immersed in NaOH, in terms of combinations of immersion temperatures of 23 °C (T_1) and 50 °C (T_2) and immersion periods of 1, 2 and 3 months (hereafter P_1, P_2, and P_3). The analyses indicate that in comparison of the four combinations of T_1P_1, T_1P_2,T_1P_3, and T_2P_1, the experimental groups (*i.e.*, nonwoven geotextiles containing 5 wt. %–20 wt. % Kevlar® fibers) have a statistically significant higher tensile strength than the control group (*i.e.*, nonwoven geotextiles without Kevlar® fibers).

Table 2. Tensile strength and statistical analyses of recycled Kevlar®/ PET/LPET nonwoven geotextiles that are immersed in NaOH, in relation to different immersion periods.

Sample Codes	Solutions	*T* (°C)	*P* (Months)	CK (wt. %)	*NS*	AV (N)	SD (N)	Scheff's Test (Kevlar® Fibers)
-	without immersion	-	-	0	5	105.82	9.51	20, 15 > 0
		-	-	5	5	113.70	6.07	
		-	-	10	5	116.32	11.53	
		-	-	15	5	117.72	11.80	
		-	-	20	5	119.14	9.95	
T_1P_1	NaOH	23	1	0	0	0	-	20, 15, 10, 5 > 0
				5	5	3.28	0.58	
				10	5	3.38	0.75	
				15	5	3.48	0.94	
				20	5	3.93	1.42	
T_1P_2	NaOH	23	2	0	0	0	-	20, 15 > 10, 5 > 0
				5	5	0.57	0.06	
				10	5	0.93	0.15	
				15	5	1.40	0.20	
				20	5	1.20	0.26	
T_1P_3	NaOH	23	3	0	0	0	-	20, 15, 10 > 5 > 0
				5	0	0	-	
				10	3	0.73	0.06	
				15	3	1.10	0.10	
				20	3	1.10	0.20	
T_2P_1	NaOH	50	1	0	0	0	-	20, 15 > 10, 5 > 0
				5	3	0.95	0.26	
				10	3	1.05	0.19	
				15	3	1.73	0.21	
				20	3	1.80	0.22	

T is the solution temperature, *P* is the immersion period, CK is the contents of Kevlar® fibers, *NS* is the number of specimens tested per sample, AV is the average values, and SD is the standard deviation. Samples are taken along the CD.

According to Scheffe's analysis, there are statistically significant differences in the tensile strength among nonwoven geotextiles with immersion conditions of T_1P_2, T_1P_3, and T_2P_1. The tensile strength of nonwoven geotextiles is ranked according to the content of Kevlar® fibers as 20, 15 wt. % > 10, 5 wt. % > 0 wt. % (T_1P_2), 20, 15, 10 wt. % > 5 wt. % > 0 wt. % (T_1P_3), and 20, 15 wt. % > 10, 5 wt. % > 0 wt. % (T_2P_1). Such results are ascribed to Kevlar® fibers that have a high modulus and high strength. The molecular chains of Kevlar® fibers have a benzene structure, and their molecular bonding has high compactness and stiffness. A value of 15 wt. % or 20 wt. % of Kevlar® fibers can effectively improve the tensile strength of nonwoven geotextiles. In addition, the majority of the fibers are arranged along the CD during the nonwoven process, and thus tensile strength along the CD is greater than that along the MD.

3.2. Effects of Immersion Durations on Tensile Strength

The effects of immersion periods in neutral, acid, and alkaline solutions on the tensile strength of recycled Kevlar®/PET/LPET nonwoven geotextiles are examined. Figure 1 and Figure 2 indicate the tensile strength of nonwoven geotextiles that are immersed in water for different periods. Thus, the tensile strength of nonwoven geotextiles does not significantly fluctuate as a result of the immersion temperatures, or of the sample types that are taken along CD or MD. Furthermore, $p > 0.05$ in one-way ANOVA analysis also indicates the differences in tensile strength are not statistically significant, indicating that the tensile strength of nonwoven fabrics is not influenced by being in a neutral environment for 4 months.

Figure 3 and Figure 4 indicate the tensile strength of nonwoven geotextiles that are immersed in a strong acid solution for 1, 2, 3, or 4 months (*i.e.*, the experimental group). The control group is the nonwoven geotextiles without immersion. Table 3indicates the tensile strength of samples that are immersed in CH_3COOH in four months. With an acid solution at 23 °C, the nonwoven geotextiles without Kevlar® fibers have a tensile strength that is in inversely proportion to the immersion periods (1, 2, 3, and 4 months), while the nonwoven geotextiles containing 5, 10, 15, or 20 wt. % Kevlar® fibers exhibit comparable tensile strength regardless of the immersion periods. However, this trend is not observed for nonwoven fabrics that are immersed in an acid solution at 50 °C, and the tensile strength slightly decreases as a result of the increasing immersion period. These differences in tensile strength of the control group and the experimental group are then further analyzed using statistical analyses. The one-way ANOVA analysis indicates that the 0/80/20 exhibits a statistically significant difference ($p < 0.05$) in tensile strength of nonwoven geotextiles that are immersed in an acid solution at 23 °C. Furthermore, when the immersion periods are considered, the post hoc Scheffe's test indicates that the significance in the 0/80/20 is ranked as 4 < 0, 1, 2, 3 (months). Namely, a 4-month immersion in an acid solution statistically significant debilitates the tensile strength of the control group, in comparison with a 0-, 1-, 2-, or 3-month immersion. In contrast, the content of Kevlar®fibers of being 10, 15, or 20 wt. % is $p > 0.05$. This shows the experimental group is acid resistant. In addition, when the experimental group is immersed in an acid solution at 50 °C, the one-way ANOVA analysis shows statistically significant differences in its acid resistance. This result suggests that the experimental group fails at remaining at its original tensile strength under an acid environment at a high temperature, which confirms to the results of the study conducted by Jeon *et al.* [17].

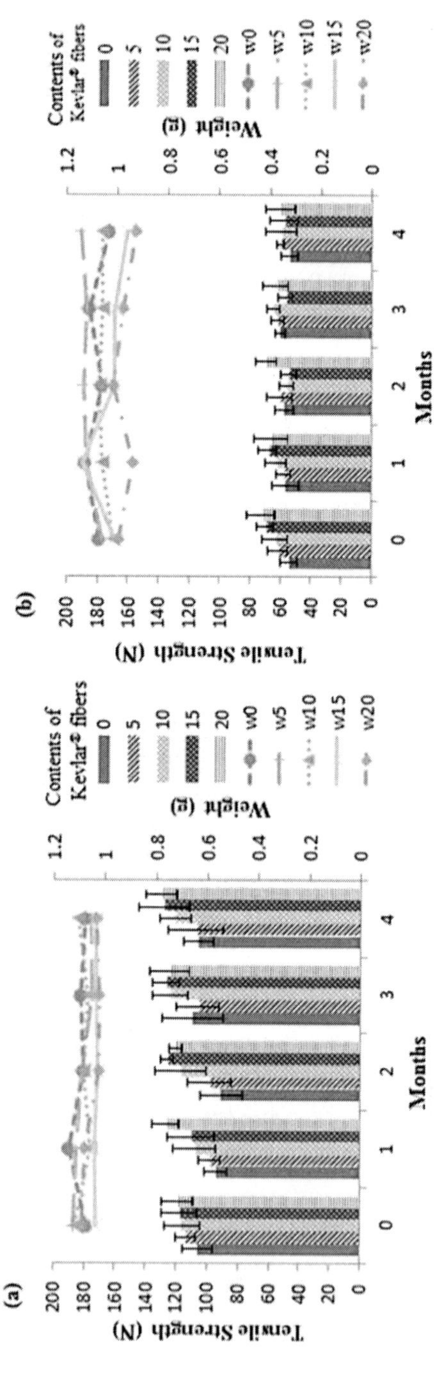

Figure 1. Tensile strength and weight along the (**a**) cross machine direction (CD) and (**b**) machine direction (MD) of recycled Kevlar®/PET/LPET nonwoven geotextiles that are immersed in H_2O at 23 °C for 4 months.

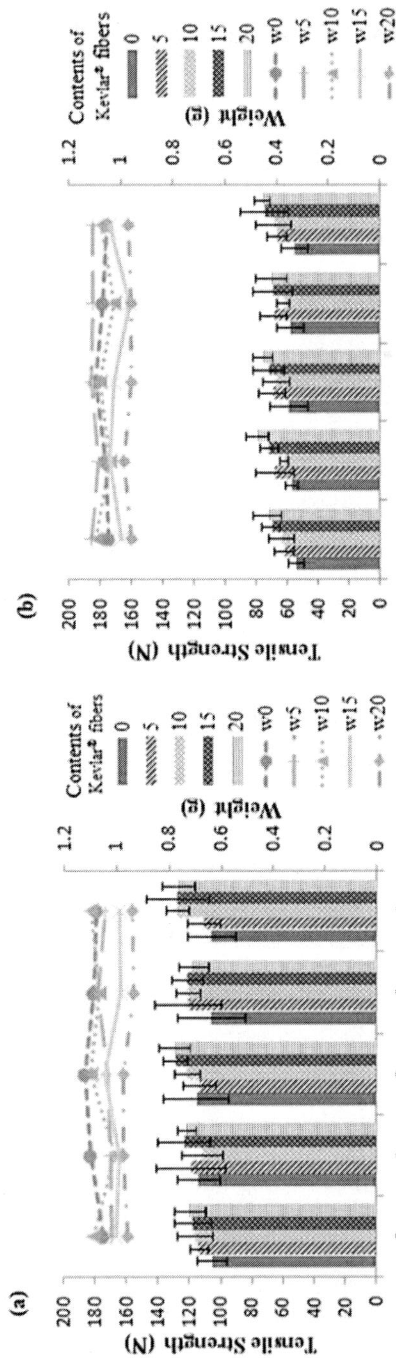

Figure 2. Tensile strength and weight along the (**a**) CD and (**b**) MD of recycled Kevlar®/PET/LPET nonwoven geotextiles that are immersed in H_2O at 50 °C for 4 months.

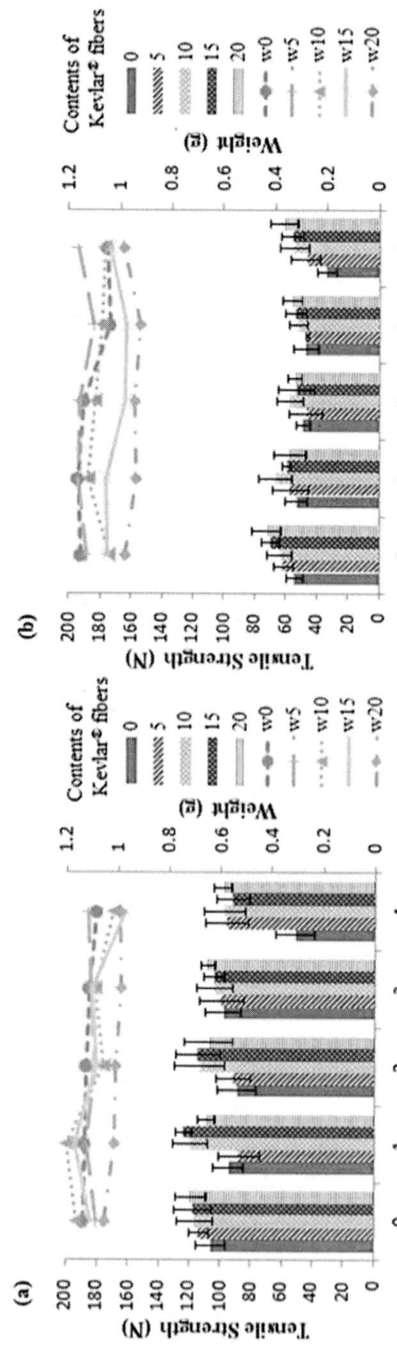

Figure 3. Tensile strength and weight along the (**a**) CD and (**b**) MD of recycled Kevlar®/PET/LPET nonwoven geotextiles that are immersed in CH_3COOH at 23 °C for 4 months.

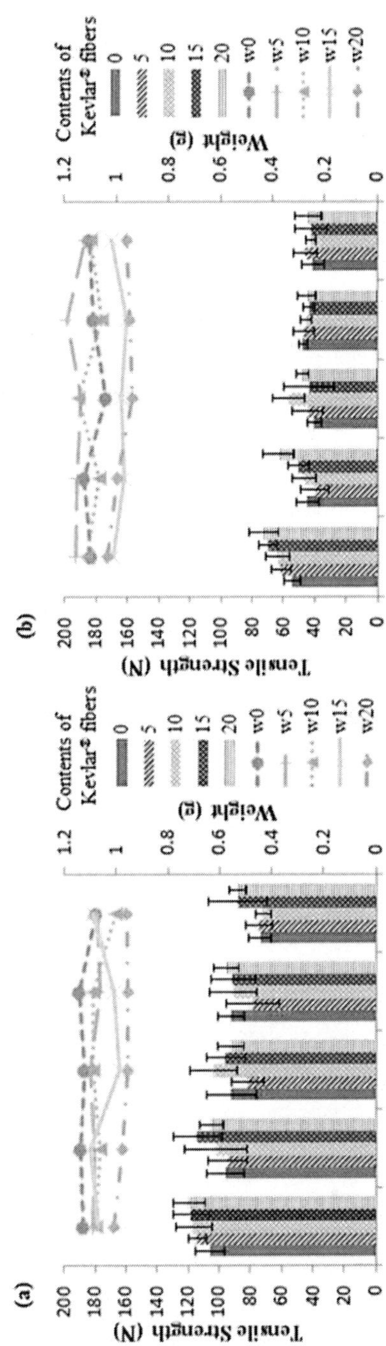

Figure 4. Tensile strength and weight along the (**a**) CD and (**b**) MD of recycled Kevlar®/PET/LPET nonwoven geotextiles that are immersed in CH_3COOH at 50 °C for 4 months.

Table 3. Tensile strength and statistical analyses of recycled Kevlar®/PET/LPET nonwoven geotextile that are immersed in CH_3COOH, in relation to different immersion periods.

CK (wt. %)	Solutions	*T* (°C)	*P* (Months)	*NS*	AV (N)	SD (N)	Scheff's Test (Immersion Period)
0	CH_3COOH	23	0	5	105.82	9.51	4 < 0, 1, 2, 3
			1	4	94.68	9.83	
			2	4	89.05	12.61	
			3	4	98.24	11.62	
			4	4	51.49	12.37	
CK (wt. %)	**Solutions**	***T* (°C)**	***P* (Months)**	***NS***	**AV (N)**	**SD (N)**	**Scheff's Test (Immersion Period)**
5	CH_3COOH	23	0	5	113.70	6.07	1 < 0
			1	4	87.75	13.60	
			2	4	91.63	10.95	
			3	4	93.53	10.16	
			4	4	90.52	10.79	
10	CH_3COOH	23	0	5	116.32	11.53	-
			1	4	119.73	11.11	
			2	4	113.33	16.28	
			3	4	103.97	11.76	
			4	4	97.69	13.61	
15	CH_3COOH	23	0	5	117.72	11.80	-
			1	4	123.53	5.29	
			2	4	114.43	14.48	
			3	4	103.80	6.73	
			4	4	92.27	10.73	
20	CH_3COOH	23	0	5	119.14	9.95	-
			1	4	108.98	5.11	
			2	4	108.00	15.91	
			3	4	108.22	4.46	
			4	4	98.62	5.82	
0	CH_3COOH	50	0	5	105.82	9.50	4 < 0
			1	4	95.83	12.34	
			2	4	91.95	16.41	
			3	4	92.12	8.39	
			4	4	73.91	7.54	
5	CH_3COOH	50	0	5	113.70	6.07	2, 3, 4 < 0
			1	4	94.40	12.24	
			2	4	81.95	10.25	
			3	4	78.60	16.74	
			4	4	74.45	8.88	
10	CH_3COOH	50	0	5	116.32	11.53	4 < 0
			1	4	101.98	19.91	
			2	4	103.73	15.32	
			3	4	90.83	15.20	
			4	4	71.93	5.03	
15	CH_3COOH	50	0	5	117.72	11.80	4 < 0
			1	4	113.98	15.81	
			2	4	95.78	12.48	
			3	4	91.30	14.34	
			4	4	88.54	18.97	
20	CH_3COOH	50	0	5	119.14	9.95	4 < 0
			1	4	105.23	7.61	
			2	4	92.30	8.34	
			3	4	95.27	7.79	
			4	4	88.26	5.59	

Samples are taken along the CD.

Figure 5 and Figure 6 show the tensile strength of the recycled Kevlar®/PET/LPET nonwoven geotextiles (experimental group) that are immersed in an alkali solution for 1, 2, 3, and 4 months. Regardless of the immersion temperatures of 23 °C or 50 °C, and regardless of the fact that samples are taken along the MD or along the CD, a 1-month immersion results in a significant decrease in the tensile strength of the experimental group. Although it is least possible for geotextiles to be used in an environment that has a pH below 13, the strong alkali solution is used in this study in order to examine the alkali resistance of recycled Kevlar®/PET/LPET nonwoven geotextiles. The control group (*i.e.*, PET/LPET nonwoven geotextile) has a complete decomposition in a NaOH solution. However, the remaining tensile strength of the experimental group is 3%–5%. This indicates that the incorporation of Kevlar® fibers improves the tensile strength of the nonwoven geotextiles. Strong alkali results in the decrease in the tensile strength of the experimental group because PET fibers undergo hydrolysis. Similarly, Mathur *et al.* state that PET is subjected to a hydrolysis as a result of being exposed to an acid or an alkali environment [15], while Bartolome *et al.* points out that a combination of water and an acid, alkali, or neutral environment renders PET with a hydrolysis that involves the depolymerization of PET to terephthalic acid (TPA), and that can be aggravated by applying a high temperature [24].

FT-IR analysis indicates that the characteristic absorption peak of PET's ester bond is C=O at a frequency of 1750–1735 cm^{-1} and C–O at 1300–1000 cm^{-1}, and the same characteristic peak is absent after the immersion in an alkali solution, as indicated in Figure 7, which shows that the ester bonds exhibit bond breaking. Cassidy, *et al.* [25] indicate that hydrolysis of ester group results in bond breaking. PET encounters alkaline hydrolysis as a result of its contact with NaOH solution. The available ester linkage on the surface is broken, and then the original polymer chain is converted to the small molecules that turn into hydroxyl and carboxylate end groups [26,27]. In addition, alkaline hydrolysis starts with pores in fibers, and the coalescence of the pores become cracks that cause PET to discompose [16,17]. As a result, the recycled Kevlar®/PET/LPET nonwoven geotextiles have a significant decrease in their weight loss, as indicated in Figure 5 and Figure 6, which is ascribed to the decrease in their tensile strength.

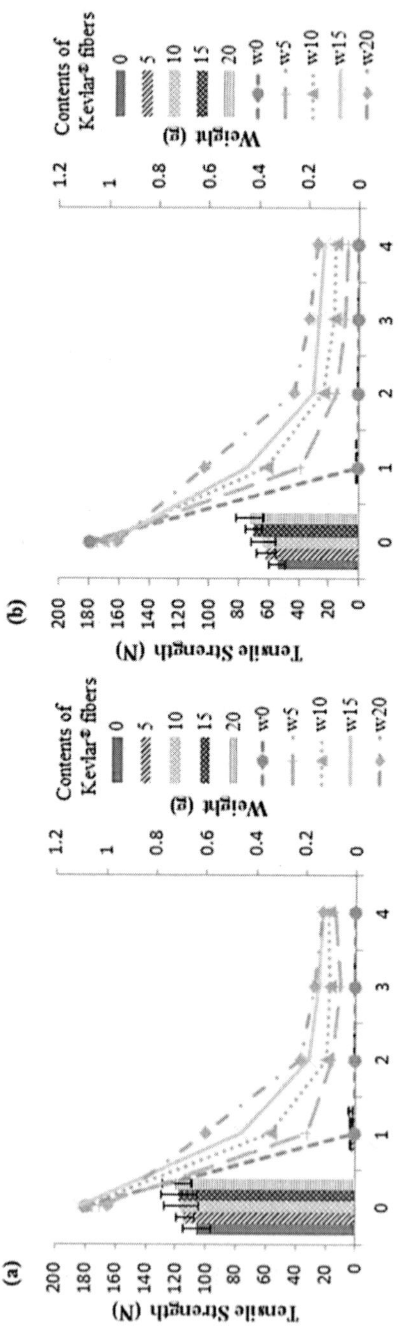

Figure 5. Tensile strength and weight along the (**a**) CD and (**b**) MD of recycled Kevlar®/PET/LPET nonwoven geotextiles that are immersed in NaOH at 23 °C for 4 months.

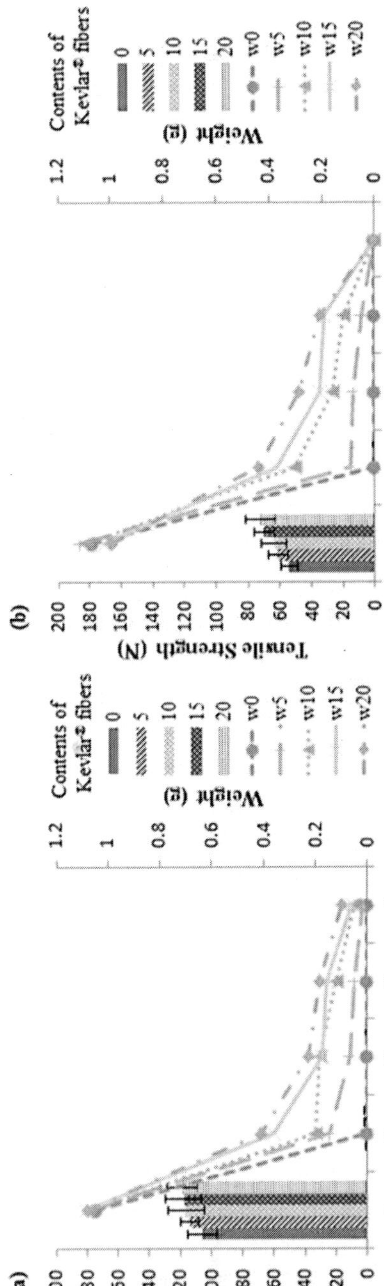

Figure 6. Tensile strength and weight along the (**a**) CD and (**b**) MD of recycled Kevlar®/PET/LPET nonwoven geotextiles that are immersed in NaOH at 50 °C for 4 months.

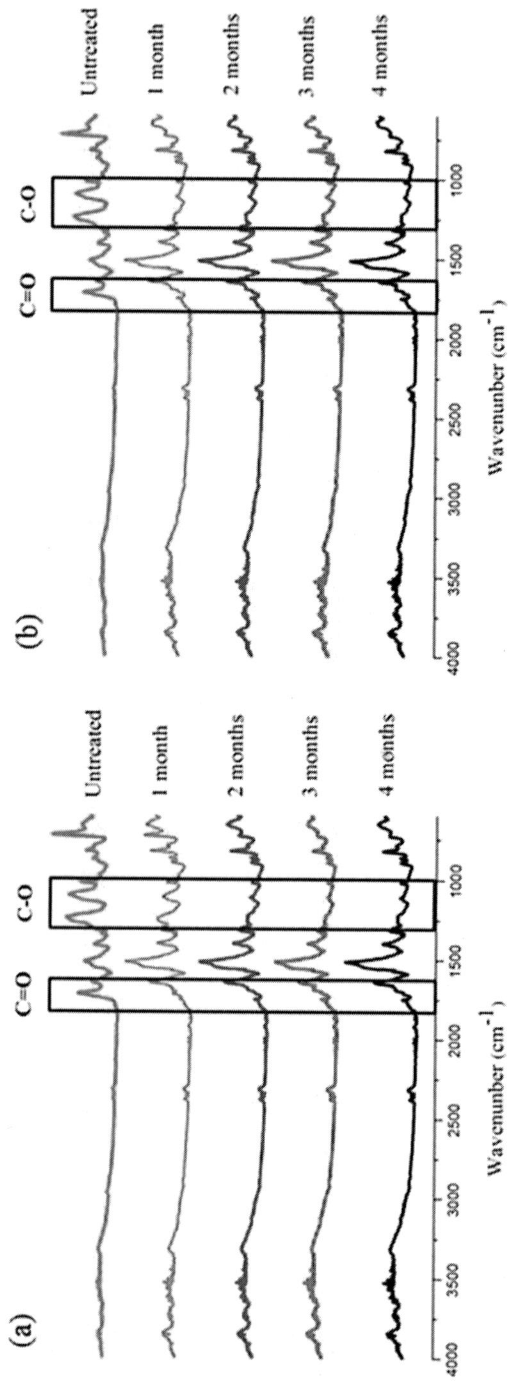

Figure 7. FT-IR spectra of nonwoven geotextiles that are immersed in NaOH at (**a**) 23 °C and (**b**) 50 °C.

3.3. Effects of Thermal Treatment on Tensile Strength

According to Figure 1, Figure 2, Figure 3, Figure 4, Figure 5, Figure 6 and Figure 7, the geotextiles have a higher fiber orientation along the CD, and samples taken along this direction are thus used for this section. Recycled Kevlar®/PET/LPET nonwoven geotextiles are composed of LPET fibers that have a sheath-core structure. The sheath has a melting point of 110 °C, based on which the thermal treatment is set to be at 120 °C. The tensile strength of nonwoven fabrics in relation to thermal treatment conditions is presented in Table 4. Regardless of the environment being neutral, acid, or alkaline, the tensile strength of the nonwoven geotextile is greater when they are thermally treated, rather than when they are not thermally treated, as indicated in Figure 8. A *t*-test is conducted in order to examine the influence of thermal treatment on the tensile response of the nonwoven geotextiles tested the tensile strength. The *t*-test results report $p < 0.05$ that there is a statistical significance as thermal treatment results in a higher tensile strength. However, the statistical analyses for NaOH solution is absent in Table 4 due to the fact that the immersion in a strong alkaline solution for 4 months results in the disappearance of tensile strength of nonwoven geotextiles. The available data is insufficient to satisfy the hypothesis for a normal distribution, and a *t*-test cannot be performed. One-way ANOVA and Scheffe's test are used to compare the tensile strength of the thermally treated and non-thermally treated geotextiles in relation to immersion periods (1, 2, 3, and 4 months) and solution temperatures (23 °C and 50 °C). For geotextiles immersed in H_2O, there is no statistical significance in the tensile strength between different immersion periods and solution temperatures. This statistical analysis is in conformity with the results in section 3.2. When applying CH_3COOH at 23 °C, the non-thermally treated geotextiles show the statistical significant difference as 1 < 0 (months), while the thermally treated geotextiles show the statistical significant difference as 2, 3, 4 < 0. In addition, the thermally treated geotextiles also have a higher alkali resistance than non-thermally treated geotextiles, as indicated in Figure 8e. Such a result is due to the constituent LPET fibers that undergo softening as a result of the thermal treatment for the nonwoven geotextiles. The fibers that are crossed are bonded by the bonding points, and thereby increase the contact areas between fibers and eventually improve the strength of nonwoven fabrics. Therefore, thermally treated nonwoven geotextiles have a higher tensile strength.

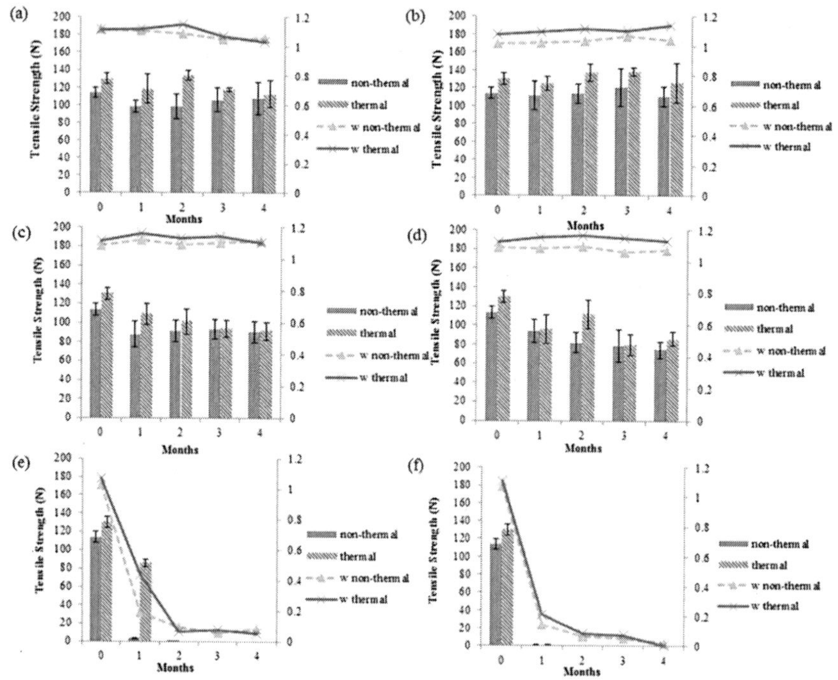

Figure 8. Tensile strength and weight along the CD of recycled Kevlar®/PET/LPET nonwoven geotextiles of (**a**) H_2O/23 °C, (**b**) H_2O/50 °C, (**c**) CH_3COOH/23 °C, (**d**) CH_3COOH/50 °C, (**e**) NaOH/23 °C, and (**f**) NaOH/50 °C, as related to the implementation of thermal treatment.

3.4. Multiple Regression Analyses of Tensile Strength

Multiple regression is used to statistically analyze the influence of the independent variables (X) on the dependent variables (Y). X refers to independent variables that are represented in $X_1, X_2, X_3\ldots, X_n$, while Y refers to dependent variable. In this study, the independent variables ($X_1, X_2, X_3\ldots, X_n$) are: Contents of Kevlar® fibers, immersion periods, and immersion temperatures. The dependent variable (Y) is the tensile strength of the nonwoven geotextiles after immersion under the specific conditions considered. Table 5 summarizes the results of multiple regression analyses, where B is the unstandardized regression coefficient that indicates the coefficient of X. β is the standardized regression coefficient that has a value between −1 ~ +1. The greater the absolute value of β, the higher the influence level of the dependent variable. The Fvalue is the result of the F test, and $p < 0.05$ suggests that the multiple regression result has statistical significance. AdjustedR^2 stands for the explanatory power of the independent variables for dependent variables. VIF value represents the collinear influence, when VIF > 5, the independent variables are highly correlated. When VIF > 10, the accuracy of multiple regression analyses is low.

Table 4. Tensile strength and statistical analyses of recycled Kevlar®/PET/LPET nonwoven geotextile that are thermally treated, in relation to different solutions and different immersion periods.

Solutions	T (°C)	P (Months)	Thermal Treatment	NS	AV (N)	SD (N)	t-test	Scheff's Test (Immersion Period)
		0	No	5	113.70	6.07		
		1	No	5	98.40	6.63		
		2	No	5	98.48	14.11		$p = 0.29$ (non-thermal)
		3	No	5	105.99	13.87		
		4	No	5	107.70	17.99		
H_2O	23						0.00	
		0	Yes	5	130.10	6.41		
		1	Yes	5	118.30	16.86		
		2	Yes	5	143.73	5.92		$p = 0.80$ (thermal)
		3	Yes	5	117.73	1.77		
		1	Yes	5	112.64	15.27		
		0	No	5	113.70	6.07		
		1	No	5	118.98	21.99		
		2	No	5	113.48	10.72		$p = 0.84$ (non-thermal)
		3	No	5	120.53	21.28		
		4	No	5	110.51	10.42		
H_2O	50						0.00	
		0	Yes	5	130.10	6.41		
		1	Yes	5	119.70	3.25		
		2	Yes	5	150.57	8.28		$p = 0.20$ (thermal)
		3	Yes	5	148.03	16.80		
		4	Yes	5	140.13	29.51		
		0	No	5	113.70	6.07		
		1	No	5	87.75	13.60		
		2	No	5	91.63	10.95		1 < 0 (non-thermal)
		3	No	5	93.53	10.16		
		4	No	5	90.52	10.79		
CH_3COOH	23						0.05	
		0	Yes	5	130.10	6.41		
		1	Yes	5	109.15	11.24		
		2	Yes	5	101.37	13.07		2, 3, 4 < 0 (thermal)
		3	Yes	5	95.46	8.66		
		4	Yes	5	94.24	9.13		
		0	No	5	113.70	6.07		
		1	No	5	94.40	12.24		
		2	No	5	81.95	10.25		1, 2, 3, 4 < 0 (non-thermal)
		3	No	5	78.60	16.74		
		4	No	5	74.45	8.88		
CH_3COOH	50						0.02	
		0	Yes	5	130.10	6.41		
		1	Yes	5	98.98	14.68		
		2	Yes	5	111.50	14.87		1, 3, 4 < 0 (thermal)
		3	Yes	5	82.94	11.05		
		4	Yes	5	86.04	7.26		

Samples are taken along the CD.

Table 5. Multiple regression analyses in the tensile strength of recycled Kevlar®/PET/LPET nonwoven geotextiles.

Model	Deposition Solution	*NS*	Variable	*B*	Beta	*F*	*p*	Adjusted R^2	VIF
1	H_2O	231	(Constant)	106.21	-	65.76	0.00	0.22	-
			Contents of Kevlar® fibers	0.98	0.47				1.00
2	CH_3COOH	196	(Constant)	113.81	-	60.14	0.00	0.48	-
			Immersion period	−6.99	−0.58				1.00
			Contents of Kevlar® fibers	0.82	0.33				1.00
			Temperature	−0.24	−0.18				1.00
3	NaOH	117	(Constant)	96.82	-	217.58	0.00	0.65	-
			Immersion period	−43.96	−0.81				1.00

The multiple regression model (1) is composed of correlations among the three designated independent variables and the tensile strength of recycled Kevlar®/PET/LPET nonwoven geotextiles that are immersed in H_2O for 1, 2, 3, and 4 months. There is no statistically significant difference in their tensile strength when samples are immersed for different lengths of time or different solution temperatures. The statistical analyses indicate that contents of Kevlar® fibers is the only influential factor of a neutral environment (H_2O) on the tensile strength of nonwoven geotextiles. The F test results demonstrate that this multiple regression analyses have statistical significance with (F (1,229) = 65.76, p = 0.00). The explanatory power is 22% without collinearity.

The multiple regression model (2) is composed of correlations among the three designated independent variables and the tensile strength of recycled Kevlar®/PET/LPET nonwoven geotextiles that are immersed in CH_3COOH for 1, 2, 3, and 4 months. The influential factors to the tensile strength of the nonwoven geotextiles that are immersed in CH_3COOH are ranked as immersion periods, contents of Kevlar® fibers, and immersion temperatures. According to the results of one-way ANOVA analyses in section 3.2, there are statistically significant differences in the tensile strength of the nonwoven geotextiles that are immersed in an acid solution for 4 months, including the five combinations of 0 wt. % of Kevlar® fibers and 23 °C, as well as 0/5/10/15/20 wt. % of Kevlar® fibers and 50 °C. In addition, the tensile strength is proportional to the contents of Kevlar® fibers. In other words, the contents of Kevlar® fibers demonstrate an influence on the tensile strength. Immersion temperature is the last independent variable that is involved in the multiple regression model (2). Because a high immersion temperature expedites the hydrolysis of PET, the tensile strength of nonwoven geotextiles is slightly lower when they are immersed in an acid solution at 50 °C, rather than when they are immersed in an acid solution at

23 °C. Based on the F test results, this multiple regression analyses have statistical significance with (F (3,192) = 60.14, p = 0.00). The explanatory power of the equation is 48% without collinearity.

The tensile strength against NaOH of recycled Kevlar®/PET/LPET nonwoven geotextiles is depicted by model (3). The analytic results indicate that the major factor is immersion periods. As a strong alkali solution with a pH = 13 is used in this study, the nonwoven geotextiles exhibit alkaline hydrolysis regardless of the immersion temperature of 23 °C or 50 °C, and is exemplified by a statistically significant decrease in the tensile strength. Based on the F test results, this multiple regression analyses have statistical significance with (F (1,115) = 217.58, p = 0.00). The explanatory power of the equation is 65% without collinearity.

4. CONCLUSIONS

This study proposes synthesizing recycled Kevlar® fibers, LPET fibers, and PET fibers to successfully create recycled Kevlar®/PET/LPET nonwoven geotextiles. First, the influences of different contents of Kevlar® fibers on the tensile strength of nonwoven geotextiles are examined. In light of experimental results and statistically analyses, the incorporation of Kevlar® fibers can improve the tensile strength of the nonwoven geotextiles. In particular, 15 wt. % and 20 wt. % of Kevlar®fibers strengthen the tensile strength by 20% to 90%, and reaches the statistical significance of $p < 0.05$. Second, the influences of immersion periods on the tensile strength of the nonwoven geotextiles are examined. The correlations between a 4-month immersion in different environment conditions and the differences in the tensile strength of the nonwoven geotextiles are 0% in a neutral environment, while there is a 30%–50% decrease for the control group and a 17%–20% decrease for the experimental group in an acid environment, and a 100% decrease for the control group and a 95%–97% decrease for the experimental group in an acid environment. Statistical analyses also confirm that the experimental group has a higher tensile strength than the control group. Third, the influences of thermal treatment on the tensile strength of the nonwoven geotextiles are examined. The thermal treatment is proved to provide a 15%–45% greater tensile strength to the nonwoven geotextiles and yields a statistical significance at $p < 0.05$. Finally, demonstrated by the multiple regression analyses, the factors that influence the tensile strength of the nonwoven geotextiles with their corresponding environments is contents of Kevlar® fibers (neutral environment), immersion periods, contents of Kevlar® fibers, and immersion temperatures (acid environment), and immersion periods (alkaline environment). The combination of recycled Kevlar® fibers and LPET fibers have been proven to significantly improve the tensile strength of nonwoven geotextiles in terms of a neutral, acid, and alkali treatment.

ACKNOWLEDGMENTS

The authors would especially like to thank Ministry of Science and Technology of Taiwan, for financially supporting this research under Contract MOST 103-2622-E-035-025-CC2.

AUTHOR CONTRIBUTIONS

In this study, the concepts and designs for the experiment, all required materials, as well as processing and assessment instrument are provided by Jia-Horng Lin and Ching-Wen Lou. Data is analyzed, and experimental results are examined by Jing-Chzi Hsieh, Chien-Teng Hsieh, Wen-Hao Hsing, Yi-Jun Pan. The experiment is conducted and the text is composed by Jia-Hsun Li.

CONFLICTS OF INTEREST

The authors declare no conflict of interest.

REFERENCES

1. Nagahara, H.; Fujiyama, T.; Ishiguro, T.; Ohta, H. Fem analysis of high airport embankment with horizontal drains.*Geotext Geomembr.* **2004**, *22*, 49–62.

2. Bouazza, A.; Freund, M.; Nahlawi, H. Water retention of nonwoven polyester geotextiles. *Polym. Test.* **2006**, *25*, 1038–1043.

3. Muthukumaran, A.E.; Ilamparuthi, K. Laboratory studies on geotextile filters as used in geotextile tube dewatering.*Geotext Geomembr.* **2006**, *24*, 210–219.

4. Lamy, E.; Lassabatere, L.; Bechet, B.; Andrieu, H. Effect of a nonwoven geotextile on solute and colloid transport in porous media under both saturated and unsaturated conditions. *Geotext Geomembr.* **2013**, *36*, 55–65.

5. Greenwood, J.H.; Myles, B. Creep and stress relaxation of geotextiles. In Proceedings of the 3rd International Conference on Geotextiles, Vienna, Austria, 7–11 April 1986; pp. 821–826.

6. Vashi, J.M.; Desai, A.K.; Solanki, C.H. Assessment of reinforced embankment on soft soil with PET and PP geotextile.*Int. J. Civ. Struct. Eng.* **2012**, *2*, 828–837.

7. Thwe, M.M.; Liao, K. Durability of bamboo-glass fiber reinforced polymer matrix hybrid composites. *Compos. Sci. Technol.* **2003**, *63*, 375–387.

8. Jawaid, M.; Khalil, H.P.S.A.; Abu Bakar, A.; Khanam, P.N. Chemical resistance, void content and tensile properties of oil palm/jute fibre reinforced polymer hybrid composites. *Mater. Des.* **2011**, *32*, 1014–1019.

9. Li, T.T.; Wang, R.; Lou, C.W.; Lin, J.Y.; Lin, J.H. Static and dynamic puncture failure behaviors of 3D needle-punched compound fabric based on weibull distribution. *Text. Res. J.* **2014**, *84*, 1903–1914.

10. Yan, R.S.; Wang, R.; Lou, C.W.; Lin, J.H. Comparison of tensile and compressive characteristics of intra/interply hybrid laminates reinforced high-density flexible foam composites. *J. Appl. Polym. Sci.* **2015**, *132*, 1–9.

11. Buggy, M.; Farragher, L.; Madden, W. Recycling of composite materials. *J. Mater. Process. Technol.* **1995**, *55*, 448–456.

12. Li, T.T.; Wang, R.; Lou, C.W.; Lin, J.H. Evaluation of high-modulus, puncture-resistance composite nonwoven fabrics by response surface methodology. *J. Ind. Text.* **2013**, *43*, 247–263.

13. Lin, C.M.; Lou, C.W.; Lin, J.H. Manufacturing and properties of fire-retardant and thermal insulation nonwoven fabrics with FR-polyester hollow fibers. *Text. Res. J.* **2009**, *79*, 993–1000.

14. Lin, J.H.; Li, T.T.; Lou, C.W. Puncture-resisting, sound-absorbing and thermal-insulating properties of polypropylene-selvages reinforced composite nonwovens. *J. Ind. Text.* **2014**, 1–13. Available online: http://jit.sagepub.com/content/early/2014/11/27/1528083714562088.full.pdf+html (accessed on 30 April 2016).

15. Mathur, A.; Netravali, A.N.; O'Rourke, T.D. Chemical aging effects on the physio-mechanical properties of polyester and polypropylene geotextiles. *Geotext Geomembr.* **1994**, *13*, 591–626.

16. East, G.C.; Rahman, M. Effect of applied stress on the alkaline hydrolysis of geotextile poly(ethylene terephthalate). Part 1: Room temperature. *Polymer* **1999**, *40*, 2281–2288.

17. Jeon, H.Y.; Cho, S.H.; Mok, M.S.; Park, Y.M.; Jang, J.W. Assessment of chemical resistance of textile geogrids manufactured with PET high-performance yarn. *Polym. Test.* **2005**, *24*, 339–345.

18. Saha, P.; Roy, D.; Manna, S.; Adhikari, B.; Sen, R.; Roy, S. Durability of transesterified jute geotextiles. *Geotext Geomembr.* **2012**, *35*, 69–75.

19. ISO 13434. Geosynthetics–Guidelines for the assessment of durability. Available online: https://www.iso.org/obp/ui/#iso:std:42713:en (accessed on 30 April 2016).

20. Lodi, P.C.; de Souza Bueno, B.; Hamada, J.; de Oliveira, E.L.; da Silva, G.H.R. Evaluation of mechanical and thermal properties after chemical degradation of PVC and HDPE geomembranes. *Electron. J. Geotech. Eng.* **2013**, *18*, 1227–1236.

21. ASTM D5322-98. Standard practice for laboratory immersion procedures for evaluating the chemical resistance of geosynthetics to liquids. Available online: http://compass.astm.org/EDIT/html_ annot.cgi?D5322+98(2009) (accessed on 30 April 2016).

22. ASTM D543-14. Standard practices for evaluating the resistance of plastics to chemical reagents. Available online: http://compass.astm.org/EDIT/html_annot.cgi?D543+14 (accessed on 30 April 2016).

23. ASTM D5035-11. Standard test method for breaking force and elongation of textile fabrics (strip method). Available online: http://compass.astm.org/EDIT/html_annot.cgi?D5035+11(2015) (accessed on 30 April 2016).

24. Bartolome, L.; Imran, M.; Cho, B.G.; Al-Masry, W.A.; Kim, D.H. Recent Developments in the Chemical Recycling of PET. In *Material*

Recycling-Trends and Perspectives; Achilias, D., Ed.; INTECH Open Access: Rijeka, Croatia, 2012; pp. 65–84.

25. Cassidy, P.E.; Mores, M.; Kerwick, D.J.; Koeck, D.J.; Verschoor, K.L.; White, D.F. Chemical resistance of geosynthetic materials. *Geotext Geomembr.* **1992**, *11*, 61–98.

26. Yang, Y.Z.; Gu, H.C. Superfine structure, physical properties, and dyeability of alkaline hydrolyzed poly(ethylene terephthalate)/silica nanocomposite fibers prepared by *in situ* polymerization. *J. Appl. Polym. Sci.* **2006**, *102*, 3691–3697.

27. Kim, E.S.; Lee, C.H.; Kim, S.H. Effects of pretreatment reagents on the hydrolysis and physical properties of PET fabrics. *J. Appl. Polym. Sci.* **2009**, *112*, 3071–3078.

CHAPTER 10

Introducing Textiles as Material of Construction of Ethanol Bioreactors

Osagie A. Osadolor *, Patrik R. Lennartsson and Mohammad J. Taherzadeh

Swedish Centre for Resource Recovery, University of Borås, SE 501-90, Borås, Sweden

ABSTRACT

The conventional materials for constructing bioreactors for ethanol production are stainless and cladded carbon steel because of the corrosive behaviour of the fermenting media. As an alternative and cheaper material of construction, a novel textile bioreactor was developed and examined. The textile, coated with several layers to withstand the pressure, resist the chemicals inside the reactor and to be gas-proof was welded to form a 30 L lab reactor. The reactor had excellent performance for fermentative production of bioethanol from sugar using baker's yeast. Experiments with temperature and mixing as process parameters were performed. No bacterial contamination was observed. Bioethanol was produced for all conditions considered with the optimum fermentation time of 15 h and ethanol yield of 0.48 g/g sucrose. The need for mixing and temperature control can be eliminated. Using a textile bioreactor at room temperature of 22 °C without mixing required 2.5 times longer retention time to produce bioethanol than at 30 °C with mixing. This will reduce the fermentation investment cost by 26% for an ethanol plant with capacity of 100,000 m^3 ethanol/y. Also, replacing one 1300 m^3 stainless steel reactor with 1300 m^3 of the textile bioreactor in this plant will reduce the fermentation investment cost by 19%.

KEYWORDS

bioethanol; fermentation; bioreactor material; textile bioreactor; reactor cost

1. INTRODUCTION

We live in a world where there is an ever-increasing demand for energy. The transportation sector accounts for a high proportion of the global energy demand, which is dominated by fossil fuels [1]. There has been growing interest in alternative fuel sources and ethanol has proven to be a viable alternative to fossil fuel in the transportation sector [2,3]. As the biofuels must compete with fossil fuels, any attempt to reduce their investment and operational costs will contribute to stimulate their consumption.

Global production of ethanol, the dominating biofuel, has increased from 50 million m^3 in 2007 to 89 million m^3 in 2013 [4], the production trends across the globe for this period are shown in Table 1. Future forecast shows that global demand for ethanol will continue to increase to an estimated value of 100 million m^3 in 2015 [5]. For 10% *w/w* ethanol production in bioreactors, this will correspond to a total fermentation volume of 785 million m^3. Despite this, the relatively cheaper price of petroleum makes some ways of ethanol production uneconomical, and it is also a hindrance to the commercial introduction of 2nd and 3rd generation ethanol into the fuel market. Several research projects have been performed on ethanol production to reduce its production costs [6].

Table 1. Global ethanol production from 2007 to 2013 by country or region in million m^3 [4].

Country	2007	2008	2009	2010	2011	2012	2013
USA	24.68	35.24	41.4	50.34	52.8	50.35	50.35
Brazil	19	24.5	24.9	26.2	21.1	21.11	23.72
Europe	2.16	2.78	3.94	4.57	4.42	4.46	5.19
China	1.84	1.9	2.05	2.05	2.1	2.1	2.63
Canada	0.8	0.9	1.1	1.35	1.75	1.7	1.98
Rest of World	1.19	1.47	3.46	3.73	2.64	2.85	4.82
World total	49.68	66.79	76.86	88.24	84.81	82.57	88.69

Ethanol is nowadays produced principally by fermentation, where the excess heat of 580 kJ/kg sugar used should be continuously released [6], and the bioreactors must be cooled [7]. In addition, the cost of the fermentation process for a conventional 100,000 m^3/y ethanol facility constitutes 11% of the total fixed capital cost of the plant [8]. In other word, the fermentation process has a large direct effect on the plant investment and operational costs [8].

A reactor is a vessel where transformation of reactants to products takes place. Reactors are generally designed using the operating conditions for the reactant to product transformation in mind, while also trying to maximise profit, ensure adequate safety and minimize environmental emissions [9]. A fermentor or a bioreactor is a reactor that provides an environment suitable for the controlled growth of a microorganism which is responsible for producing a product of interest [10,11]. A bioreactor should be made of materials that are inert and do not facilitate the development of unwanted microorganisms [12]. It should provide adequate temperature control, operate well under sterilization conditions (with chemicals or temperature) [11], provide good contact area for the microbes and the substrate [9], have adequate charging inlet and discharging outlets, have a means of adequate sampling [11], and provide adequate time for the desired product to be produced [9]. Most conventional bioreactors are designed to have very low surface area to volume ratio, which increases the cooling requirements of the bioreactor [11].

This paper introduces a novel bioreactor for producing bioethanol made from textiles. The textile bioreactor has the potential for higher flexibility in ethanol production. Its lower cost compared to stainless steel, could lead to a reduction of the cost of producing bioethanol, thereby making investments in the ethanol market more attractive.

2. RESULTS AND DISCUSSION

The textile bioreactor used in this work is a novel bioreactor for producing bioethanol. It is made from a backbone of textile which is coated with several layers of polymers to make it resistant to chemicals, gas and liquid leakage. It is flexible, long lasting and can withstand temperatures up to 150 °C. Some of the advantages of using a textile bioreactor for bioethanol production include: it does not corrode, it can withstand the tough environmental conditions encountered during fermentation, it is a far cheaper alternative than the currently used bioethanol bioreactors, it is light, and designed for easy and safe transportation, installation and operation, and it is ultraviolet irradiations (UV) resistant, it can be sterilized with steam at

121 °C and with chemicals. It was originally developed for biogas production [13], but it was never examined for any other fermentation products. In this work, this new textile bioreactor was developed for bioethanol production, its performance was examined and the results are presented here.

2.1. Textile vs. Other Materials for Construction of Bioreactors

The materials used for constructing bioreactors must be able to withstand the physiochemical conditions encountered while running the bioreactor and during clean-up and sterilization [14]. Apart from stainless steel, other materials that could be used for making bioreactors include carbon steel, borosilicate glass, polytetrafluoroethylene (PTFE) plastic, and ceramics [14]. Only stainless steel 304 and to a lesser extent reinforced carbon steel are currently being used to make industrial ethanol bioreactors. The other materials are normally added to stainless steel bioreactors at specific points for certain purposes (e.g., borosilicate glass used in sight glasses) [14].

Bioethanol is produced by fermentation. Bioethanol fermentation takes place under slightly acidic conditions (pH between 4 and 6), temperatures ranging from 25 to 38 °C, generally without oxygen, and in a liquid medium. Ensuring that only the desired microorganism is what grows in the bioreactor is necessary to ensure the fermentable sugars are converted to the product of interest [15]. It is essential that the material used for constructing bioreactors for producing bioethanol does not affect the fermentation process and can be sterilized when needed. For all the experiments performed in the textile bioreactor, it was autoclaved for sterilization at 121 °C for 20 min and 2 bar pressure. This created a sterile working condition for the textile bioreactor. There were no incidences of bacterial contamination in all the experiments performed in the textile bioreactor, as there were no areas for harbouring unwanted microorganisms, which is one of the main reasons why stainless steel is used as the current material for making bioreactors [16]. The material of construction of the textile bioreactor has been proven to resist diverse environmental conditions (pH 3–12) [13]. In addition the material when burnt does not ignite, but rather forms a semi-solid composite which recoils inward, making the bioreactor fire resistant. The material was designed to have high tensile strength with high flexibility to make its assembly and disassembly easy. Table 2shows some advantages and disadvantages of using certain materials of construction for ethanol bioreactors. Considering these features and the comparison in Table 2, as a bioreactor material of construction, the textile bioreactor is an excellent choice for bioethanol production.

Table 2. Advantages and disadvantages of possible materials for construction for ethanol bioreactors [14,17,18].

Material	Modification	Advantage	Disadvantage
Textile	Layered with some polymers and UV filter	Portable. Corrosion proof. Good sterility. More cost effective than stainless steel. Can withstand high temperature. Leak proof. Long life span. Can be designed to have regions that are transparent, for easy process monitoring.	Currently a horizontal vessel.
Stainless steel 304	–	Cheapest of all the stainless steel. Leak proof. Good sterility. Can withstand high temperature and pressure. Corrosion proof. Long life span.	Quite expensive.
Carbon steel	Reinforced with stainless steel	Cheaper than 304 stainless steel. Leak proof.	Corrosion and contamination.
Borosilicate glass	–	Transparent. Inert to chemicals.	Very fragile.
Plastic	–	Very portable. Cheap.	Leaks and short life span. High chances of contamination.
Ceramics	–	Chemically stable. Wear resistant.	Brittle. Prone to thermal shock.

2.2. Reactor Cost Comparison

A major challenge facing biofuel production is its economic feasibility [19]. Bioethanol production consists of the collection of feedstock, pre-treatment of feedstock (if the feedstock is starch or lignocellulosic based), fermentation, distillation and possibly dehydration [20]. The fermentation cost of a 100,000 m^3/y ethanol production facility contributes 11% of the total plant cost, while the bioreactor cost makes up 32% of the fermentation cost [8]. In this section a comparison is made between the investment cost of stainless steel bioreactors and textile bioreactors excluding operation cost (maintenance and installation cost). Typically, the installed cost (investment and operation cost) of a stainless steel reactor is 1.7 times its purchase cost [21], while that of a textile bioreactor is 1.5 times its purchase cost [13].

The purchase cost for a 1000 m^3 textile bioreactor is $100,000. Table 3 shows the purchasing cost of different reactor sizes, for both the developed textile bioreactor and stainless steel reactors. The purchasing cost of stainless steel reactor was estimated using Equations (1) and (2) (see Section

3.5). For all reactor volumes considered, the purchasing cost of the developed textile bioreactor was far less than half the purchasing cost of the stainless steel bioreactor. Considering a 100,000 m^3/y ethanol production facility using sucrose as its raw material and having a fermentation time between 10 and 15 h [22,23], this plant will require a bioreactor volume between 1000 and 1500 m^3 for the fermentation only. If this plant has just one 1300 m^3 stainless steel bioreactor, replacing this with a 1300 m^3 textile bioreactor will reduce the fermentation investment cost by 19%, and the total plant investment cost of the facility by 2.1%.

Table 3. Purchasing cost of developed textile bioreactors and 304 stainless steel reactors.

Reactor Size (m^3)	Purchasing Cost of Developed Textile Bioreactor ($)	Purchasing Cost of 304 Stainless Steel Reactor ($)
500	66,000	201,000
1,000	100,000	282,000
1,300	130,000	325,000
1,500	150,000	352,000

2.3. Mixing and Temperature Control in the Textile Bioreactor

The mass transfer in a bioreactor affects the net productivity of the system [11,24]. The two crucial aspects of mass transfer in a bioreactor are the uniform distribution of the product and substrate in the bulk liquid, and the transfer of substrate into the cells and the products out of the cells. Mixing helps to minimize local variation of concentration and temperature in a bioreactor [11]. Mixing in liquid media can be achieved by agitation, or with the aid of a stirrer, or by the use of a pump for recirculation, depending on the viscosity of the liquid and if it media is single- or multi-phased [25]. For the textile bioreactor, mixing was performed using a recirculation pump. To determine the effectiveness of the mixing in the bioreactor and the possibility of it being used for continuous fermentation, experiments were performed where samples were collected from the sampling point at the centre of the textile bioreactor and from the exit pipes from the bioreactor. One of the basics of a continuous stirred tank reactor (CSTR) is having same concentration in the reactor as what leaves the reactor. FromFigure 1, it is clearly observed that there is no significance difference between the concentration in the bioreactor (samples from the centre of the reactor) and that leaving the bioreactor (samples from the exit pipe). This shows that there is the possibility of the textile bioreactor being used for batch, fed-batch, and continuous fermentation. In addition to ensuring uniform substrate and product distribution in the textile bioreactor, the mixing also helped to provide a good transfer of substrate into and products out of the yeast (Figure 1), as the sugar was fully consumed about the same time as peak ethanol concentration was reached. Thus the mixing by recirculation in the textile bioreactor is effective.

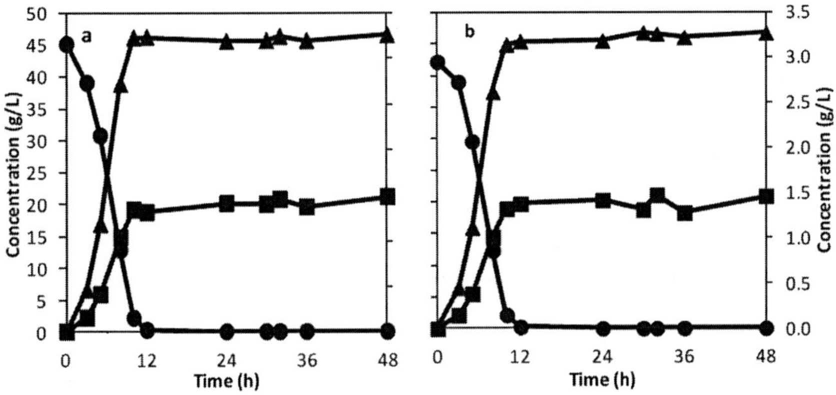

Figure 1. Concentration of sucrose (●) and ethanol (■) on the primary axis (left side), and glycerol (▲) on the secondary axis (right side), with samples taken from the exit pipe (**a**) and centre of the reactor (**b**) with time, to determine the effectiveness of mixing by recirculation.

Temperature control is essential for optimal product formation as every microorganism has a temperature range in which it functions optimally. For anaerobic condition that for *S. cerevisiae* is around 30 °C [15]. For lab scale production heat is normally added to the system while for the large industrial bioreactors with low surface to volume ratio, cooling is necessary [11]. Because of the nature of the material used for the developed textile bioreactor, cooling can easily be achieved by recirculation of chilled water, while heating can be achieved with hot water. The area to volume ratio of a 1000 m^3 textile bioreactor is 0.96, while that of a conventional 1000 m^3 bioreactor having a height to diameter ratio of 3 it is 0.62. The higher area to volume ratio of the textile bioreactor makes cooling (or heating) easily achievable because the heat loss (or gained) by evaporation and radiation increases with increasing area to volume ratio. Temperature control was achieved in the textile bioreactor as described in Section 3.2. In addition, the recirculation of the fluid also helped to ensure temperature uniformity in every part of the reactor. Figure 2 and Figure 3 shows the result of the experiments performed with temperature control, while Figure 4 shows the result of the experiments without temperature control. For both cases, ethanol yields greater than 87% of the theoretical values were reached, while the experiments where temperature was maintained at 30 °C had higher fermentation rates and peak product concentrations were reached in less than 24 h. This shows that the temperature control developed for the textile bioreactor is effective.

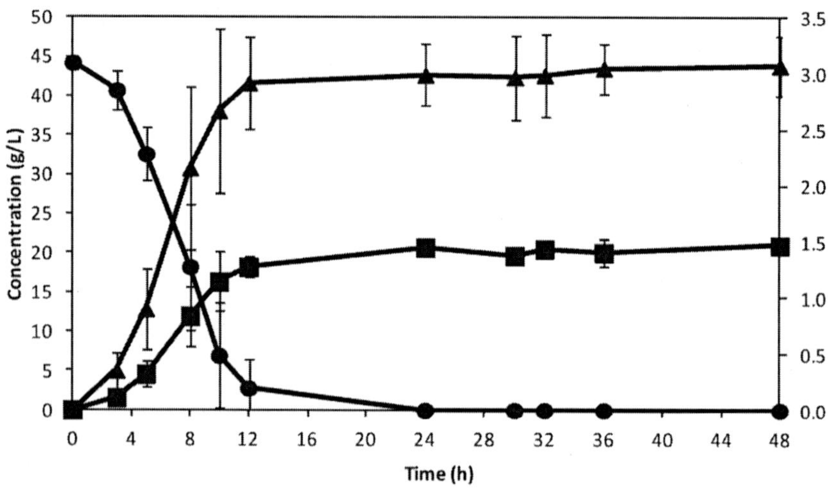

Figure 2. Concentration of sucrose (●) and ethanol (■) on the primary axis, and glycerol (▲) on the secondary axis with time, at 30 °C and with mixing.

2.4. Fermentation in the Textile Bioreactor and Its Economics

To determine how effective the textile bioreactor was for producing bioethanol, lab scale experiments were performed under different operating conditions. Temperature and mixing were varied, while the pH of all the experiments performed was around 6.0 ± 0.2. Figure 2 shows the result of the experiment performed in the textile bioreactor at 30 °C and with mixing. For this experiment, the yield of ethanol from the experiment (using the initial sucrose concentration of 44.2 g/L) was 0.48 ± 0.01 g/g, which is 88% of the theoretical value, and it took an average of 15 h for the yeast to consume the sugar. Comparing this fermentation time with that from a similar work were 10 g/L of yeast was used and it required a fermentation time of 10 h [22], shows that the fermentation time is good. Thus fermentation takes place effectively well and at a good rate in the textile bioreactor. From Figure 2, the average of the peak ethanol concentration was 20.04 ± 0.53 g/L, using the average fermentation time gave the specific productivity to be 1.34 ± 0.02 g $L^{-1}{\cdot}h^{-1}$.

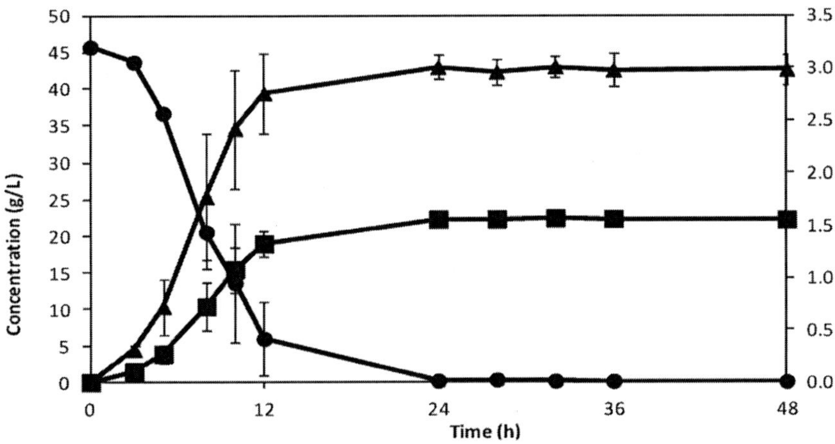

Figure 3. Concentration of sucrose (●) and ethanol (■) on the primary axis, and glycerol (▲) on the secondary axis with time, at 30 °C without mixing.

Figure 3 shows the result of the experiment where temperature was fixed at 30 °C without mixing. The yield of ethanol from the experiment was 0.49 ± 0.01 g/g, and it took an average of 20 h for the yeast to consume the sugar. Comparing Figure 2 and Figure 3 shows that mixing did not affect the fermentation rate that much when temperature is held at 30 °C without mixing, as the specific productivity for this case was 1.04 ± 0.01 g L^{-1} h^{-1} in comparison to 1.34 ± 0.02 g L^{-1} h^{-1} with mixing. To produce the same amount of ethanol as that which is produced when mixing is controlled; the textile bioreactor volume used has to be 1.29 times the one used with mixing. Taking a 1000 m^3 bioreactor operating with mixing and temperature control, this bioreactor will cost $282,000 but a 1300 m^3 textile bioreactor will cost $130,000 (see Table 3). In addition, the cost of agitation and mixing in a bioreactor accounts for 24% of the fermentation cost of a 100,000 m^3 ethanol/y production facility [8]. Using a textile bioreactor operated at 30 °C without mixing can eliminate the need for the cost of agitation and mixing, and it gives a bioreactor cost reduction of $152,000.

Figure 4 shows the result of the experiment that was performed at room temperature of 22 °C with and without mixing. The ethanol yield from the experiment with mixing was 0.49 ± 0.01 g/g, and it took an average of 40 h for the yeast to consume the sugar, while that without mixing had an ethanol yield of 0.49 ± 0.02 g/g, and it took an average of 42 h for peak ethanol concentration to be reached. Mixing did not affect the fermentation rate that much as the specific productivity with mixing was 0.55 ± 0.01 g L^{-1} h^{-1} while that without mixing was 0.53 ± 0.02 g L^{-1} h^{-1}. This result shows that there is a possibility of running the textile bioreactor without temperature control and mixing. The slower fermentation rate from producing bioethanol at 22 °C can be accommodated by increasing the retention time and the bioreactor volume. ComparingFigure 2 and Figure 4, the same amount of ethanol per hour will be produced in both cases if the volume of the textile bioreactor for the production without temperature control and mixing is 2.53 times that with temperature control and mixing. Taking a 1000 m^3 bioreactor with temperature control and mixing, the purchasing cost of a 1000 m^3 bioreactor is $282,000 while that of a 2530 m^3 textile bioreactor (consisting of one 1000 m^3 reactor and one 1500 m^3 reactor) is $250,000 (see Table 3). In addition, operating the textile bioreactor without temperature control and mixing also reduces the total fermentation cost, as the cost of temperature control, mixing and agitation accounts for 26% of the fermentation cost in a 100,000 m^3/y ethanol production plant [8]. For the same bioethanol production rate it is more economical to use a larger volume of the textile reactor without temperature control and mixing than a smaller bioreactor volume with temperature control, mixing and agitation.

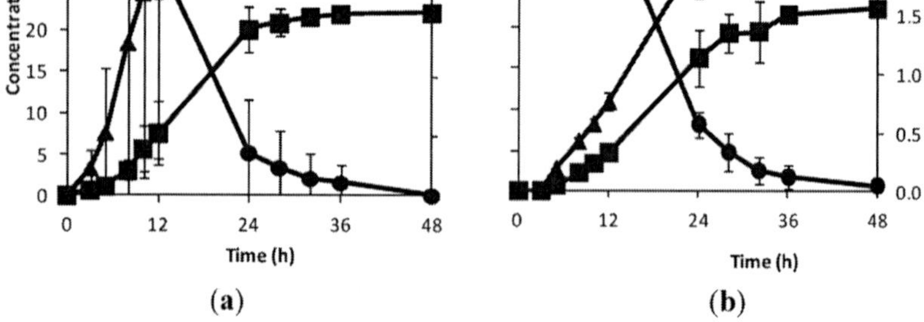

Figure 4. Concentration of sucrose (●) and ethanol (■) on the primary axis, and glycerol (▲) on the secondary axis with time, at room temperature of 22 °C (**a**) with and (**b**) without mixing.

Comparing the scenario where there is mixing but the textile bioreactor is operated at 22 °C, the size of the reactor in this case would be 2.44 times the size of that operated at 30 °C. Using a 1000 m^3 bioreactor at 30 °C will cost $282,000 while a 2440 m^3 textile bioreactor running at 22 °C will cost $250,000. However running the textile bioreactor at 22 °C with mixing will only reduce the fermentation cost by 2%, which is not as economical as 26% cost reduction obtained by running it at 22 °C without mixing [8].

For all experiments performed in the textile bioreactor there were no incidences of bacterial contamination. From the results, experiments performed at 30 °C had faster fermentation rates than the ones performed at room temperature. For a continuous process, both temperature control and mixing will be essential to achieve high dilution rate. The results of the experiments show that the textile bioreactor can be used for bioethanol production at different conditions of temperature and mixing.

3. EXPERIMENTAL SECTION

3.1. Microorganism

Dry ethanol red yeast (*Saccharomyces cerevisiae*) from Fermentis (Strasbourg, France) was used for the fermentation. A starting concentration of 1 g/L of the dry yeast was used.

3.2. Textile Bioreactor

A schematic of the lab scale prototype of the developed bioreactor is shown in Figure 5. The material of construction (MOC) was a textile backbone coated with several layers to protect against pressure, chemicals in the reactor, weather conditions and to be gas proof. The material was developed by FOV Fabrics AB (Borås, Sweden) primarily for biogas reactors and it was welded to form a reactor by Kungsäter Industri AB (Kungsäter, Sweden). The lab scale bioreactor had a total volume of 30 L and a working volume of 25 L. The dimensions of the bioreactor were 100 cm length, 50 cm breadth, and 6 cm width. It had an opening of 4 cm diameter, which serves for sample collection; probe stand, thermometer stand, and gas exit. The dimensions of the outlet and inlet tapered from 9 to 4 cm at the bioreactor entrance, to allow for easy loading of the bioreactor. The dimension of the tubes connected to the inlet and outlet were 8 mm.

The means for temperature control was made using Poly(vinyl chloride) (PVC) tubing of 50 m length and a woollen blanket; the PVC tubing was connected to a GD120 grant thermostatic circulator (GD Grant Instruments Ltd., Cambridge, UK). The tubing was wound 14 times, covering the whole perimeter of the textile bioreactor at the bottom, as such,*Saccharomyces cerevisiae* was not exposed to thermal shocks [12]. The tubing and the textile bioreactor were enclosed by the woollen blanket. The temperature of the thermostatic circulator was set at 33 °C. A 200 rpm Watson Marlow compact peristaltic pump (W-M Alitea AB, Stockholm, Sweden) was used for recirculation of the fluid in the reactor for mixing.

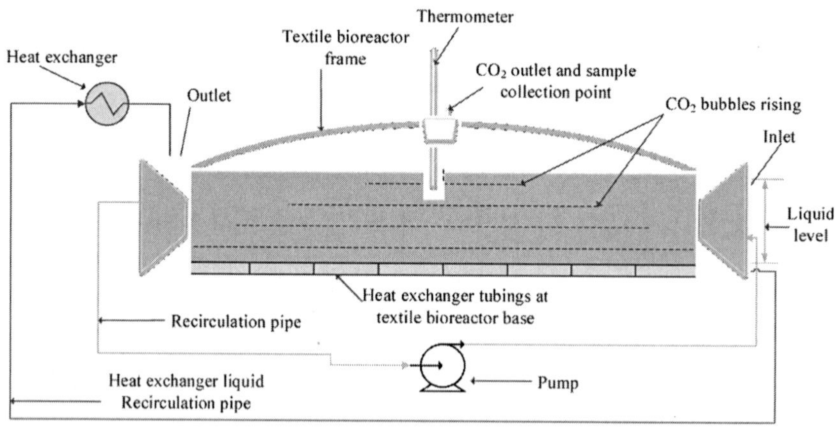

Figure 5. Schematic diagram of the textile bioreactor lab scale prototype setup.

3.3. Experimental setup

Experiments with temperature (30 °C and room temperature of 22 °C) and (with and without) mixing using recirculation as process variables were carried out in the textile bioreactor. Sucrose (50 g/L) was used as the carbon and energy sources, supplemented with 7.5 g/L $(NH_4)_2SO_4$, 3.5 g/L KH_2PO_4, 0.75 g/L $MgSO_4{\cdot}7H_2O$ and 1.0 g/L yeast extract. The total liquid volume in the reactor was 25 L for all experiments. The flow rate used for recirculation of the fluid for the experiments with mixing was 0.924 L/min. Sucrose concentration dropped between 44 and 47 g/L when the feed stream used for each experiment was autoclaved. Each experiment was performed in duplicate.

To determine the effectiveness of the mixing in the bioreactor, samples were taken from the recirculation pipe and from the centre of the reactor.

3.4. Analytical Method

Liquid samples from the textile bioreactor were analysed using high-performance liquid chromatography (HPLC) with a hydrogen-based ion exchange column (Aminex HPX-87H, Bio-Rad, Hercules, CA, USA) at 60 °C and 5 mM 0.6 mL/min H_2SO_4 eluent. A refractive index detector (Waters 2414, Waters Corporation, Milford, MA, USA) and a UV detector (Waters 2487) were used with the HPLC. The samples used for the HPLC analysis were centrifuged for 10 min at 10,000 ×*g* and the liquid portion stored for analysis. The samples were stored at −20 °C prior to HPLC analysis. All reported error bars and interval represents two standard deviations. The yield for each experiment was calculated using the concentrations measured by the HPLC after autoclaving.

3.5. Cost Estimation

The purchasing cost of 500, 1000, 1300 and 1500 m^3 versions of the developed textile bioreactors were provided by FOV Fabrics AB, and compared with those of 304 stainless steel reactors. The purchasing cost of a 304 steel reactor was estimated using Equation (1) [21], where V is the reactor volume in gallons and F_m is 2.4 for 304 stainless steel [21]. The Chemical Engineering Plant Cost Index (CEPCI) as at when Equation (1) was developed is 325.8 [21]. The capital cost was then updated to January 2014 values using Chemical Engineering Plant Cost Index (CEPCI) of 572.8 [26]. The updated cost was computed using Equation (2):

$$C = F_m \exp[11.662 - 0.6104(\ln V) + 0.04536(\ln V)^2] \quad (1)$$

$$C_{updated} = C\,(I_{updated}/I) \quad (2)$$

4. CONCLUSIONS

In this work, a novel bioreactor for bioethanol production was introduced. The bioreactor has textile as its core material of construction. For the experiments performed on a lab scale prototype textile bioreactor, the optimum result for possible continuous production of bioethanol was that obtained from the experiment performed at 30 °C and with mixing, having a yield of 0.48 ± 0.01 g/g and it took an average of 15 h for all the sugar to be fermented and peak bioethanol production level to be reached. For the same ethanol production rate, the need for mixing and temperature control can be eliminated by using a textile bioreactor 2.5 times the volume of that needed with temperature and mixing control. Doing this For a 100,000 m^3/y bioethanol production facility will reduce the fermentation investment cost by 26%, while replacing a 1300 m^3 stainless steel reactor with a 1300 m^3 textile bioreactor running at 30 °C and with mixing will reduce the fermentation investment cost by 19% and the total plant investment cost by 2.1%.

ACKNOWLEDGMENTS

The textile bioreactor material used for this work was kindly supplied by FOV Fabrics AB, Borås, Sweden. The authors will like to appreciate Ramkumar Nair and Mostafa Jabbari for assisting in making the heat exchanger for the textile bioreactor.

AUTHOR CONTRIBUTIONS

Osagie A. Osadolor did the experimental design and performed all the experiments, wrote most of the manuscript and was responsible for part of the idea. Patrik R. Lennartsson was responsible for part of the idea and manuscript. Mohammad J. Taherzadeh was responsible for part of the idea and manuscript. All authors have given their approval to the final version of the manuscript.

CONFLICTS OF INTEREST

The authors declare no conflict of interest.

REFERENCES

1. Wyman, C.E. Ethanol production from lignocellulosic biomass: Overview. In *Handbook of Bioethanol: Production and Ultilization*; Wyman, C.E., Ed.; CRC Press: Boca Raton, FL, USA, 1996; pp. 1–3.

2. Balat, M.; Balat, H. Recent trends in global production and utilization of bio-ethanol fuel. *Appl. Energy* **2009**, *86*, 2273–2282.

3. Americano da Costa, M.; Normey-Rico, J.E. Modeling, control and optimization of ethanol fermentation process.*World Congr.* **2011**, *18*, 10609–10614.

4. Licht, F.O. Ethanol Industry Outlook 2008–2013 Reports. 2014. Available online: http://www.ethanolrfa.org/pages/annual-industry-outlook (accessed on 23 July 2014).

5. Sarkar, N.; Ghosh, S.K.; Bannerjee, S.; Aikat, K. Bioethanol production from agricultural wastes: An overview. *Renew. Energy* **2012**, *37*, 19–27.

6. Palacios-Bereche, R.; Ensinas, A.; Modesto, M.; Nebra, S.A. New alternatives for the fermentation process in the ethanol production from sugarcane: Extractive and low temperature fermentation. *Energy* **2014**, *70*, 595–604.

7. Magazoni, F.; Monteiro, J.B.; Colle, S.; Cardemil, J.M. Cooling of ethanol fermentation process using absorption chillers. *Int. J. Thermodyn.* **2010**, *13*, 111–118.

8. Maiorella, B.L.; Blanch, H.W.; Wilke, C.R.; Wyman, C.E. Economic evaluation of alternative ethanol fermentation processes. *Biotechnol. Bioeng.* **2009**, *104*, 419–443.

9. Hess, W.T. *Kirk Othmer Encyclopedia of Chemical Technology*; John Wiley & Sons: Hoboken, NJ, USA, 1995.

10. Leib, T.M.; Pereira, C.J.; Villadsen, J. Bioreactors: A chemical engineering perspective. *Chem. Eng. Sci.* **2001**, *56*, 5485–5497.

11. Blakebrough, N. Fundamentals of fermenter design. *Pure Appl. Chem.* **1973**, *36*, 305–316.

12. Benz, G.T. Bioreactor design for chemical engineers. *Chem. Eng. Prog.* **2011**, *107*, 21–26.

13. Rajendran, K.; Aslanzadeh, S.; Johansson, K.; Taherzadeh, M.J. Experimental and economical evaluation of a novel biogas digester. *Energy Convers. Manag.* **2013**, *74*, 183–191.

14. Olisti, Y. Build better industrial bioreactors. *Chem. Eng. Prog.* **1992**, *1992*, 55–58.

15. Fakruddin, M.; Islam, M.A.; Ahmed, M.M.; Chowdhury, N. Process optimization of bioethanol production by stress tolerant yeasts isolated from agro-industrial waste. *Int. J. Renew. Sustain. Energy* **2013**, *2*, 133–139.

16. Santos, N.M.; Amorim, J.; Nascente, P.A.P.; Freire, C.M.A.; Cruz, N.C.; Rangel, E.C. Enhancement of corrosion resistance AISI 304 steel by plasma polymerized thin films. *Mater. Res. Soc. Symp. Proc.* **2013**, *1499*.

17. Ashby, M.F.; Jones, D.R.H. *Engineering Materials 1: An Introduction to Properties, Applications and Design*; Elsevier: Amsterdam, The Netherlands, 2012.

18. Baddoo, N.R. Stainless steel in construction: A review of research, applications, challenges and opportunities. *J. Construct. Steel Res.* **2008**, *64*, 1199–1206.

19. Berni, M.; Dorileo, I.L.; Prado, J.M.; Forster-Carneiro, T.; Meireles, M.A.A. Advances in biofuel production. In *Biofuels Production*; Scrivener Publishing LLC: Beverly, MA, USA, 2013; pp. 11–58.

20. Taherzadeh, M.J.; Lennartsson, P.R.; Teichert, O.; Nordholm, H. Bioethanol production processes. In *Biofuels Production*; John Wiley & Sons: Hoboken, NJ, USA, 2013; pp. 211–253.

21. James, R.; Couper, W.R.P.; Fair, J.R.; Walas, S.M. Costs of individual equipment. In *Chemical Process Equipment*; Elsevier Inc.: Amsterdam, The Netherlands, 2005; pp. 663–669.

22. Basso, L.C.; Basso, T.O.; Rocha, S.N. Ethanol production in Brazil: The industrial process and its impact on yeast fermentation. In *Biofuel*

Production—Recent Developments and Prospects; InTech: Rijeka, Croatia, 2011; pp. 85–100.

23. Richard, T.; Elander, V.L.P. Ethanol from corn: Technology and economics. In *Handbook on Bioethanol: Production and Utilization*; Wyman, C.E., Ed.; CRC press: Boca Raton, FL, USA, 1996.

24. Shupe, A.M.; Liu, S. Effect of agitation rate on ethanol production from sugar maple hemicellulosic hydrolysate by Pichia stipitis. *Appl. Biochem. Biotechnol.* **2012**, *168*, 29–36.

25. Paul, E.L.; Atiemo-Obeng, V.; Kresta, S.M. *Handbook of Industrial Mixing: Science and Practice*; John Wiley & Sons: Hoboken, NJ, USA, 2004.

26. Mignard, D. Correlating the chemical engineering plant cost index with macro-economic indicators. *Chem. Eng. Res. Des.* **2014**, *92*, 285–294.

CHAPTER 11

Development of plasma enhanced antiviral surgical gown for healthcare workers

Venkatesan Parthasarathi1* and Govindharajan Thilagavathi2

1 Department of Fashion Technology, PSG College of Technology, Peelamedu, Coimbatore 641004, India.

2 Department of Textile Technology, PSG College of Technology, Peelamedu, Coimbatore 641004, India.

ABSTRACT

Plasma treatments are gaining popularity in the textile industry due to their numerous advantages over conventional wet processing techniques. In this study, nonwoven fabrics spunbond polypropylene, polyester basis weight of 25 g/m^2 and microporous PTFE film were used to develop tri-laminate antiviral surgical gown. The outer layer of spunbond polypropylene was treated with plasma enhanced fluorocarbon. The plasma treated polypropylene was also treated with titanium nano dispersion finish in a single bath pad-dry-cure method. The pore sizes of nonwoven fabrics were characterized using tri-nocular microscope. The titanium nano finished polypropylene was characterized by scanning electron microscope. The tri-laminate surgical gown was developed with outer layer of plasma treated polypropylene, middle layer of PTFE and inner layer of polyester nonwoven. Liquid barrier properties of surgical performance were analyzed by viral penetration, antibacterial, spray impact penetration, hydrostatic resistance, tensile properties and moisture vapour permeability to estimate their suitability for antiviral surgical gown. It is observed that plasma treated gown shows 99.04% bacterial reduction as compared to untreated, thus providing barrier against microbes. The plasma treatment does not alter the weight and tensile properties of surgical gown. The developed plasma enhanced flurocarbon treated tri-laminate surgical gown offer sufficient liquid barrier properties for level 4 protection as per Association for the Advancement of Medical Instrumentation classification. Moisture vapour transfer rate of plasma treated tri-laminate gown decreased by 21% in comparison with untreated nonwoven gown.

KEYWORDS

Viral penetration, Hydrostatic, Moisture vapour, Impact penetration, Antibacterial

1. INTRODUCTION

Surgical gown should be made out of liquid proof fabric to protect the blood-borne infectious microbes from penetrating through the fabric. Risks of patients are contamination from both endogenous and exogenous microorganisms and risk of healthcare workers are contamination from various blood-borne pathogens due to occupational exposure to patient blood and body fluids (Unsal et al. 2005). Surgical gowns should provide an effective protective barrier against the transfer of microorganisms, particulates and fluids, in addition to acting as an aseptic barrier for the patient's protection, in order to minimize strike-through and the potential for personnel contamination (Rutala and Weber 2001). Due to the prevalence of HIV and hepatitis B and C viruses in the patient population, the barrier efficacy of protective surgical gowns have gained importance. During surgery, in an operating room, a patient's blood can penetrate surgical gown material and possibly contaminate the surgeon's skin, if not well protected. Several blood borne pathogens have the potential to spread in this manner, the most important being the HIV and the hepatitis B virus, which are related to AIDS and hepatitis (Loveday et al. 2007).

Surgical gown materials should have antimicrobial properties and blood repellency properties in order to protect patients from contamination by surgical staff during operations and also to protect the surgical team from infectious blood and other body fluids (Mews 2009). Commercial available polyester, polyester-cotton and cotton plain woven fabrics were treated with fluropolymer and found that polyester and cotton twill fabrics have high protection (Midha et al. 2014). To apply both antimicrobial and flurochemical repellent finishes to nonwoven surgical gown fabrics, a one-bath process was investigated by Huang and Karen (2007). Etching of fabrics with functional groups from a gas based plasma treatment can impart functionality, such as grafting polar surface groups to impart hydrophilicity on a hydrophobic fibre surface. There have even been studies conducted on greige fabric treated with plasma and research was made to allow greige fabric to be dyed without the requirement of fabric preparation such as desizing, scouring, and bleaching. The research found that exposing a greige fabric to plasma altered the surface so that it became hydrophilic. Through the formation of polar groups on the fabric surface, during plasma treatment, the increas hydrophilicity was studied for use in pretreating textiles (Gawish et al. 2008; Rajpreet et al. 2004).

Flurochemical finishes, commonly referred to as flurocarbon finishes, are the most widely used repellent finish in the textile industry and both natural and synthetic fibres can be treated (Huang and Leonas 1999; Jessie et al. 2007). During laundering, dry cleaning, and tumble drying excellent chemical and thermal stability of flurocarbons allow them to have great durability. In addition, flurocarbon finishes can be applied at a lower add-on than any other repellent finishes (Tyner 2007). A new standard (ASTM F2407, Specification for Surgical Gowns Intended for Use in Healthcare Facilities) which facilitates the gown selection process was developed under the auspices of the American National Standards Institute (ANSI) and the Association for the Advancement of Medical Instrumentation (Association for the Advancement of Medical Instrumentation 2003) responded to these shortcomings. A barrier material's performance is classified into four levels by this standard. Table 1 shows Classification of barrier performance and the results of the following tests as per Association for the Advancement of Medical Instrumentation (AAMI).

Table 1. Classification of barrier performance as per AAMI

AAMI barrier performance	Test	Result
Level 1	AATCC42:2000 Impact penetration test	≤4.5 g
Level 2	AATCC42:2000 Impact penetration test	≤1.0 g
	AATCC 127:1998 Hydrostatic pressure test	≥20 cm
Level 3	AATCC42:2000 Impact penetration test	≤1.0 g
	AATCC 127:1998 Hydrostatic pressure test	≥50 cm
Level 4	ASTM F1671:2003 Viral penetration test	Pass

This research work aims to develop plasma enhanced fluorocarbon treated surgical gown to protect the healthcare workers from pathogens spreading from patients as well as reverse contamination. Liquid barrier properties of surgical performance were analyzed by viral penetration, antibacterial, spray impact penetration, hydrostatic resistance, tensile properties and moisture vapour permeability.

2. METHODS

Spun bond polypropylene and polyester nonwoven fabrics were obtained from Mogul nonwoven industry, Turkey. The diameter of the individual polypropylene filament was approximately 0.2 microns and diameter of the individual polyester filament was 0.8 microns. The basis weights of the both nonwoven fabrics were 25 grams per square meter. The prime components of this research are polypropylene and polyester nonwoven fabrics. As polypropylene fibre has stereo-regular isotactic molecular structure, due to its high degree of crystallinity, good handling, strength and a high enough melting point for normal use, it is being used in many industrial applications, as well as in healthcare protective clothing. It has non polar and hydrophobic nature which is a good liquid repellent property. Considering its liquid barrier protection PP nonwoven is used as an outer layer in the developed surgical gown. Polyester nonwoven fabric is used as an inner layer in the developed surgical gown because of its wickability characteristics. The wickability nature of polyester nonwoven transfer the moisture from inner to outer layer through capillary action which makes the wearer comfort.

Figure 1 shows the plasma experiments conducted on the samples. Prototype downstream Atmospheric Pressure Plasma Reactor is equipped with a controlled stage and a solid state power generator. The stage was cooled using a chiller. The monomer was pumped into a lab designed heated evaporator through an equipped tube. The fluorocarbon gas used to provide hydrophobicity was C_3F_6. The radio frequency required to build the electromagnetic field was 600 watts. The electrode applicator height was 3 mm. The flow rate monomer was maintained as 0.6 ml/min and the exposure time was 10 seconds.

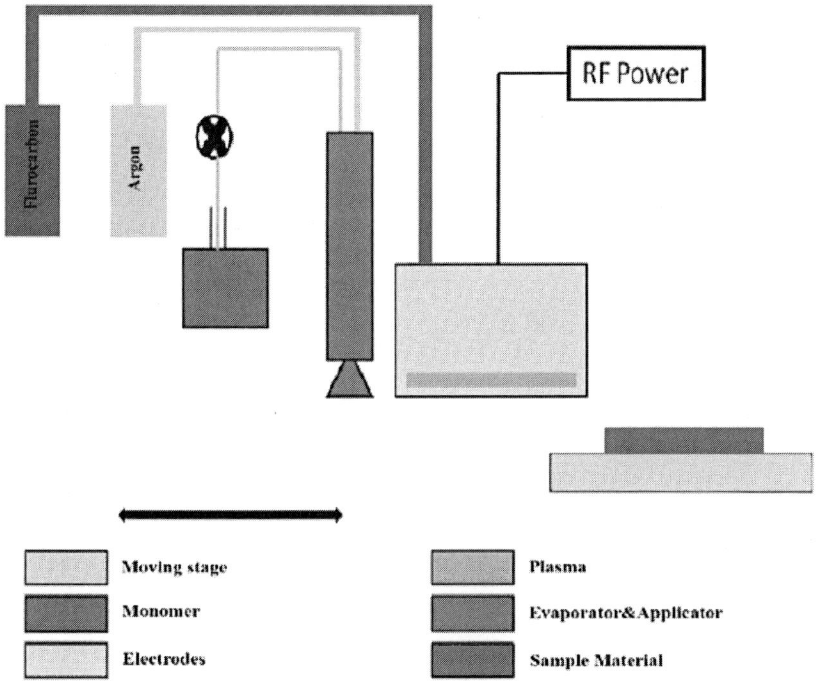

Figure 1. Plasma experiment setup.

Fabric was cut into 15 cm square samples. The samples were ironed to reduce wrinkling and to give an even surface height for treatment. Samples were then mounted on the stage using double sided sticky tape. Sample surface was made as smooth as possible, and loose yarns were removed or taped down. Once stable plasma was generated, stage was turned on to move sample into the plasma field for the desired exposure time for the pre-deposition surface activation treatment. After the sample passed out of the plasma field, it passed beneath the applicator, which deposited the monomer to condense and graft activated species on the fabric surface. Monomer flow was then turned off. Stage direction was reversed, and the sample passed through the plasma field for second time to induce free radical polymerization of the monomer on the fabric surface. Fabric sample was then removed from the stage and the treatment was done on both the sides.

A schematic of the flurocarbon plasma reaction mechanism is shown in Figure 2, in which I^+ refers to an ion, I* refers to a hot neutrals, the dashed lines represent energy transfer through the polymer and the curved lines represent species diffusion through the polymer. Through the deposition of a flurocarbon as an overlayer, etching of surface proceeds. $C_xF_y$radicals are the precursors to polymerize deposition followed by ion activation of surface sites. The polymer layer is formed by energetic ion sputtering and F atom etching. The polymer layer is the main inhibitor for the transport of species and delivery of activation energy to the surface. The oxygen reacts with the flurocarbon species to release etch product such as COFx. Finally by this reaction lower surface energy to the polypropylene nonwoven surface was imparted.

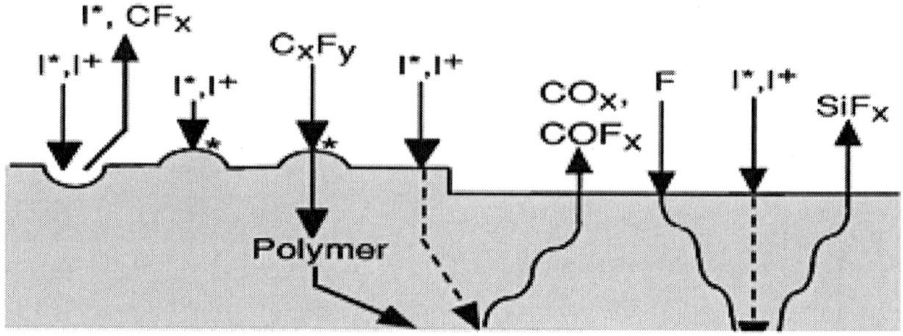

Figure 2. Flurocarbon plasma mechanism.

For the application of synthesized Titanium nano dispersion, the most widely used method of application pad-dry-cure process was used. The plasma treated polypropylene nonwoven fabric was coated with 1% nano dispersion with material to liquor ratio of 1:20 and 0.5% anionic binder. The fabric samples were immersed in the bath followed by padding through squeezed rollers at a pressure of 147 N/cm^2 to remove the excess liquid. Each fabric sample was padded with the solution twice to ensure even distribution of solution. After padding, the fabric was dried naturally and then cured at 150°C for 2 min. The wet pick up was 89%. The add-on of the titanium nanoparticles was 87%. Pore size of the plasma enhanced nano finished outer nonwoven fabric was characterized using tri-nocular microscope.

The plasma enhanced fluorocarbon treated fabric samples were characterized using scanning electron microscope. Tri-laminate gown with plasma enhanced flurocarbon treated nano finished polypropylene as an outer layer, PTFE film as a middle layer and polyester nonwoven fabric as an inner layer were bonded together using a fusing machine at a temperature of 165°C with pressure of 98 N/cm^2. The developed plasma treated surgical gown was tested and analyzed for the following properties such as Viral penetration analysis, antibacterial, tensile, Spray impact penetration, hydrostatic resistance and moisture vapour permeability.

Risks of healthcare workers are contamination from various blood-borne pathogens due to occupational exposure to patients' blood and body fluids. So the antibacterial testing was performed for the developed surgical gown. Light activate antibacterial mechanism which involves interaction with and transferring energy to oxygen so as to provide a source of an active or excited oxygen species known as singlet oxygen. Singlet oxygen is an activated form of the usual triplet state molecular oxygen. Upon exposure to normal light, the dyes used in the present invention generate singlet oxygen that kills microorganisms. Bacteria require less light though they are more sensitive to the singlet oxygen generated by the light activated dye. Surgical room illumination of about 2000 foot candles is more than sufficient. Test swatches of this plasma treated nonwoven were tested for microbiological kill by inoculating the swatches.

The antimicrobial substrate so made comprises of a light activatable singlet oxygen generating substance to inhibit the growth of bacteria selected from the group consisting of Staphylococcus aureus, Pseudomonas aeruginosa, Escherichia coli, Mycobacterium bovis, methicillin resistant Staphylococcus aureus and Proteus vulgaris. The bacterial concentration was 0.5%.

Swatches with a 0.5% concentration of bacteria were exposed to light of 2000 foot candles or dark for a given period of time. After exposure for the required time, the number of live bacteria remaining on the test swatches was determined. The log of the remaining live bacteria versus the exposure time to light of 2000 foot candles and dark were determined. After a suitable incubation, bacterial colonies were counted, percent kill calculated. Recovery procedure started with transfer of samples into a sterilized capped tube containing about a dozen 3 mm glass beads and 5 ml of sterile phosphate buffered saline. After sealing the cap, the tube was shaken vigorously for one minute and then placed for 30 seconds in a low intensity ultrasonic cleaning bath. One ml of the recovery fluid was also placed directly in a 100 mm Petri plate. For one to two days the plates were incubated at 30°C., until countable. The zero time counts recovered from the swatches were compared with the inoculum count. To obtain and average log zero time reading, all of the zero time contents were combined by averaging their base 10 logarithms. Log reduction was calculated as the difference between the samples and average zero time logarithms. Percent kill was calculated as $100(1- Log_{10})$.

Viral penetration testing of surgical gown using ASTM F 1671 was carried out to investigate the antiviral property. ASTM F1671 is the test method used to measure the resistance of materials used in protective clothing for penetration by blood borne pathogens, using bacteriophage under the condition of continuous liquid contact. The test system has been designed for measuring penetration of surrogate microbe for hepatitis B virus (HBV), hepatitis C virus (HCV) and HIV. The protective clothing materials to be tested are intended to provide protection against blood, body fluids and other potentially infectious materials. The surface tension range for blood and body fluids is approximately 42 – 60 dynes/cm. In order to simulate the wetting characteristics of blood and body fluids, the surface tension of the ΦX174 bacteriophage suspension is adjusted to approximate lower end of this surface tension range (42 dynes/cm).

The ΦX174 bacteriophage was selected as the most appropriate surrogate for the blood borne pathogens mentioned because it satisfies all of these criterias. The ΦX174 bacteriophage is a non-enveloped 25 – 27 nm virus (similarly to HCV, the smallest pathogen) with an icosahedral or nearly spherical morphology and excellent environmental stability. It is non-infectious to humans, has a limit of detection which approaches a single virus particle, grows rapidly, and can be cultivated to reach high titers similar to HBV (the most concentrated pathogen mentioned). Test samples are prepared by randomly cutting the protective clothing material into approximately 75×75 mm swatches. Test samples are exposed to approximately 60 ml of the ΦX174 bacteriophage suspension. At the conclusion of the test, the observed side of the test sample is rinsed with a sterile assay medium and then analyzed for the presence of the ΦX174 bacteriophage. Surgical protective clothing ‘pass/fail’ determinations are based on detection of penetration.

3. RESULTS AND DISCUSSIONS

3.1 Scanning Electron Microscope (SEM) analysis of plasma enhanced flurocarbon treated polypropylene nonwoven fabric

The SEM observation of untreated and treated samples as shown in Figures 3 and 4, showed small scale erosion and fragmentation surface, which can be seen as small particles. It was noticed that control sample had a smooth surface. Flurocarbon plasma treated polypropylene nonwoven showed some redeposited particles of flurocarbon etched away during plasma treatment, due to oxidation of polypropylene plasma induced free radical polymerization of the monomer. Riikka Vaananen et al. (2010) obtained similar results. They noticed that the control sample had a smooth surface and the plasma treated sample showed some redeposited particles.

Figure 3. SEM image of untreated polypropylene.

Figure 4. SEM image of plasma treated polypropylene.

3.2 Pore size characterization of plasma finished fabric

The average pore size diameter of the polypropylene plasma treated and untreated fabrics are presented in Table 2. The pore size of flurocarbon treated outer layer polypropylene nonwoven fabric is 0.25 μm. The inner layer polyester nonwoven fabric pore size is 0.35 μm. The *t*-test p value (0.001) showed a significant difference in the mean pore diameter of untreated and plasma treated polypropylene nonwoven fabric.

A surface reaction mechanism of flurocarbon acrylate based repellents react through the C=C at the end of the chain. The electrons involved in the carbon-double bond-carbon are open to attack by a free radical. These free radicals are supplied by the plasma treatment, and exist on the activated fabric surface and are deposited following the monomer deposition. Once opened, free radical polymerization ensues. The graft polymerization occurs when the monomer reacts with a free radical on the surface of the fabric. The free radical attacks the double bond of the monomer, opening the bond and forming a new bond between the fabric and the monomer, through reaction which reduces the pore size of treated polypropylene nonwoven.

Table 2. Pore size of plasma enhanced flurocarbon treated nonwoven fabric

Nonwoven fabric	Pore size diameter (μm)	Maximum perimeter (μm)	Minimum perimeter (μm)
Untreated polypropylene	0.29	1.356	0.942
Plasma treated polypropylene	0.25	1.248	0.921
Polyester	0.35	1.471	0.985

3.3 Viral penetration analysis

Table 3 shows the liquid barrier characteristics of the surgical gowns. The developed tri-laminate fabric with plasma treated polypropylene as an outer layer, PTFE film as a middle layer and polyester nonwoven fabric as an inner layer has passed the viral penetration test against the hepatitis B and hepatitis C, human Immunodeficiency viruses and surrogate ΦX174 bacteriophage. Plasma treated polypropylene single layer surgical gown failed the viral analysis due to the absence of PTFE film as it is not a barrier against viruses. The PTFE film is constructed from stretched polytetrafluroethylene which prevents the penetration of liquid but the interconnected pore structure allows the transmission of water vapor. PTFE film transports water molecules by molecular diffusion according to the sorption-diffusion model through the polymer matrix using difference in partial water-vapor pressure across the film as the driving force. Dual laminated flurocarbon treated polypropylene and polyester surgical gown failed in viral penetration analysis as it is observed that test liquid penetrated at the opposite of the fabric due to absence of PTFE film. In the absence of PTFE in the developed single layer plasma treated polypropylene gown, it also failed in the viral penetration analysis.

3.4 Antibacterial activity

Table 4 shows the antibacterial log reduction and percentage reduction of microorganism in titanium nano composite flurocarbon finished polypropylene nonwoven fabric with exposure to light of 2000 foot candles and no light. When plasma is applied to the substrate, the free electrons break the chemical bonds, upon collision with substrate, creating free radical on the surface which influence adhesion of titanium nano composites. The graph in Figure 5 shows that there is a significant reduction in bacterial count and its average is 99.04% when exposed to light, whereas in zero time contact and no light, there is no reduction in bacterial count. The titanium nano dispersion photocatalytic activity kills bacteria, when it is illuminated by light with energy higher than its band gaps and the electrons in titanium dioxide jump from the valence bond to the conduction bond.

Table 3. Viral penetration analysis of plasma enhanced flurocarbon treated gowns

S.No	Surgical gown	Viral penetration analysis
1	Plasma treated Polypropylene surgical gown	Fail
2	Plasma treated Polypropylene with PTFE dual layer surgical gown	Pass
3	Tri-laminate Plasma surgical gown	Pass
4	Dual laminated Plasma treated Polypropylene and polyester gown	Fail

Table 4. Antibacterial log reduction of plasma enhanced flurocarbon treated fabric

Samples			Bacterial count log value	
	Zero time	No light	Light (2000 fc)	% Antibacterial reduction
Control	2.4	2.3	2.3	24.20
Sample 1	2.4	2.3	1	98.94
Sample 2	2.4	2.4	0.9	99.08
Sample 3	2.4	2.4	0.9	99.08
Sample 4	2.4	2.3	1	98.94
Sample 5	2.4	2.3	0.9	99.08

Light: 2000 foot candels.
Time: 60 minutes.

The flurocarbon treated polypropylene has slightly higher antibacterial resistance than untreated but it further enhances the property. The outer layer of polypropylene has control bacteria growth due to flurocarbon treatment. The current results are similar with the studies of Souma (2012), who observed that repellency is high for plasma treated sample and high concentration of flurocarbon is required to increase repellency.

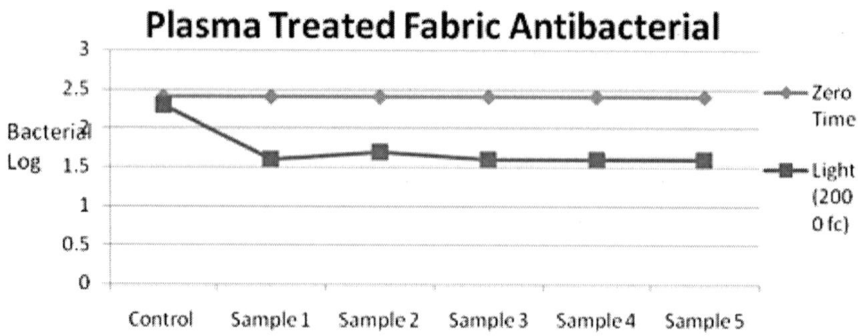

Figure 5. Plasma enhanced flurocarbon treated fabric antibacterial reduction.

3.5 Tensile and elongation of plasma enhanced flurocarbon treated surgical gown

Plasma treatment is a surface phenomenon and hence it should not significantly affect the tensile strength of nonwoven fabric. Research on atmospheric pressure plasma treated nonwovens shows that tensile properties are dependent on the severity of the treatment and the inherent properties of fabric such as the degree of entanglement of fibres, fibre-to-fibre frictional contact, laying of fibers in the web structure and the individual strength of fibers. Table 5 shows the results of tensile strength of the untreated and plasma treated fabrics. No significant changes were observed in results of tensile strength after plasma treatment. Plasma treated fabric had only slight decrease in tensile strength wherever the fabric had higher energy per area.

Table 5. Tensile and tearing strength of untreated and plasma enhanced flurocarbon treated surgical gown fabrics

Sample	Tensile strength (N)		Tearing strength (N)	
	MD	CD	DI	DII
Untreated Polypropylene (Single layer) surgical gown	105	62	10	5
Plasma treated Polypropylene (Single layer) surgical gown	98	54	7	3
Plasma treated Polypropylene with PTFE dual layer surgical gown	122	61	16	7
Tri-laminate Plasma surgical gown	149	74	21	13

MD - Machine direction, CD - Cross direction, DI - Direction 1, DII - Direction 2.

The tensile strength of the tri-laminate gown in machine direction is greater than the cross direction of the sample as the laying of polypropylene fibre during extrusion is in the machine direction. Paired *t*-test analysis of the tensile strength in machine and cross direction shows that there is no significant difference at 95% confidence interval. This result is supported by findings of Vaananen et al. (2010). They reported that the plasma treated nonwoven fabric was lower in tensile strength of both machine and cross direction as compared to untreated nonwoven.

3.6 Impact penetration and hydrostatic resistance of plasma enhanced flurocarbon treated surgical gown

The impact penetration and hydrostatic resistance characteristics of untreated, plasma treated single layer polypropylene, dual laminate surgical gown and tri-laminate are shown in the Figure 6. It is observed that the weight of water penetrating through the fabric during impact penetration test decreases as the weight of fabric increases. The results are statistically significant for untreated fabric, plasma treated single layer, dual-laminate gown and tri-laminate gown at 95% confidence level. Application of plasma finish using flurocarbon reduces the surface energy of the fabric and does not permit the water or other fluids to absorb and spread on the fabric surface. Higher fabric weight offers more resistance to water penetration because of higher fabric thickness and more number of fibers per unit area of the fabric. More number of fibers per unit area lead to higher entanglements and compact fabric structure, resulting in the reduction of water penetration.

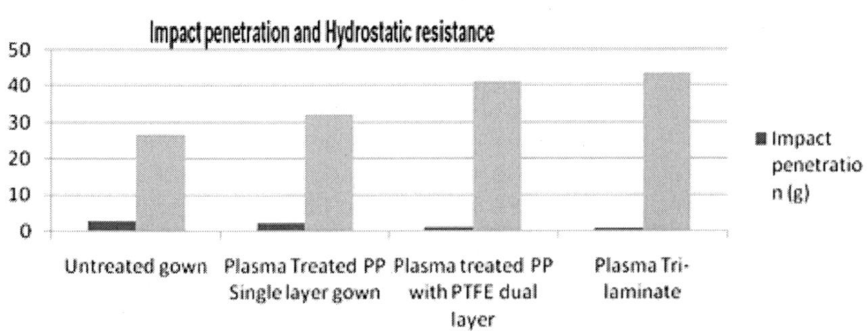

Figure 6. Impact penetration and hydrostatic resistance of surgical gown.

The hydrostatic resistance increases with increase in weight of the fabric in all gowns. The hydrostatic results are statistically significant for untreated nonwoven fabric, plasma treated single layer and dual layer fabrics at 95% confidence level. Plasma treatment with flurocarbon imparts lower surface energies to nonwoven fabric surface which increases resistance of treated fabric against blood. Dual layered of plasma treated polypropylene with PTFE gown offers better resistance than Single layer plasma treated polypropylene gown because of presence of PTFE.

According to the AAMI barrier performance classification, plasma treated tri-laminate surgical gown can be used for level 2 protection. The flurocarbon treatment makes the surface of fabric hydrophilic due to etching of fabric by deposition of a flurocarbon monomer following ion activation of surface sites. The polymer surface layer is the main inhibitor for developing repellency in nature. The present results are comparable with those of Rajpreet et al. (2004) who also performed hydrostatic resistance of flurocarbon treated nonwoven. Plasma treated nonwoven fabric is showing better hydrostatic resistance than untreated.

3.7 Moisture vapour permeability of plasma enhanced flurocarbon treated surgical gown

Moisture vapour transfer rate (MVTR) is the ability of a fabric to transfer perspiration in the form of moisture vapour through it. The MVTR of untreated and plasma treated sample were shown in Figure 7. MVTR of plasma treated tri-laminate gown decreased by 21% in comparison with untreated nonwoven gown. As the layer of nonwoven increases, water vapour permeability reduces. Moisture repellency is higher in flurocarbon treated nonwoven due to etching of fabric by deposition of a flurocarbon following ion activation of surface sites. The polymer surface layer is the main inhibitor for developing repellency in nature. Hence the moisture vapour transfer of flurocarbon treated gown is lower as compared to other gowns. The study of Souma (2012) supports that the flurocarbon treatment improves repellency whereas reduces the moisture vapour transfer of nonwoven fabric.

Test of significance of means using paired *t*-test showed that the means of plasma treated and untreated fabrics were significantly different at 95% confidence interval. A negative correlation between the water vapour permeability against the increasing thickness of fabric has also been noticed. The regression equations as $Y = -986.36\times + 1139.09$.

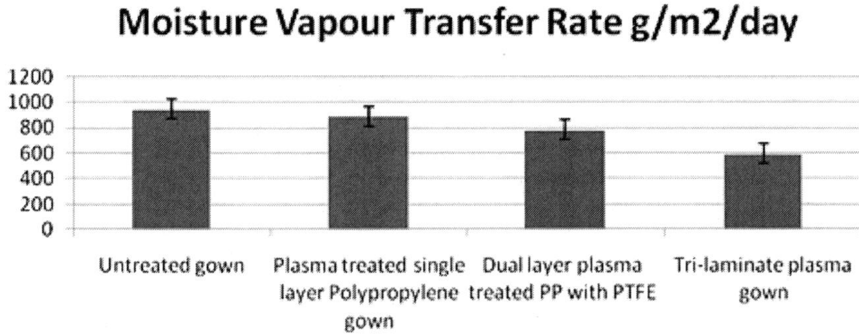

Figure 7. Moisture vapour permeability of untreated and plasma enhanced flurocarbon treated surgical gown.

4. CONCLUSIONS

Plasma enhanced fluorocarbon treated single, dual and tri-laminate surgical gowns have been used to study their suitability for antiviral surgical gowns. Pore size of plasma treated polypropylene nonwoven is smaller than untreated polypropylene, thereby influencing antibacterial resistance. Both the tri-laminate and dual layer plasma treated PTFE gowns passed in viral penetration analysis whereas plasma treated single layer surgical gown and dual laminated plasma polypropylene with polyester failed in viral penetration analysis due to absence of PTFE film. Impact penetration of Plasma treated single and dual layer gown offers level 1 protection and plasma tri-laminate gown has level 2 protection. Hydrostatic resistance of plasma treated tri-laminate gown offers level 2 protection. Tri-laminate plasma gown has better impact and hydrostatic resistance as compared single and dual laminate gowns. MVTR of plasma treated tri-laminate gown decreased by 21% in comparison with untreated nonwoven gown. The developed plasma enhanced fluorocarbon treated surgical confirms level 4 protection according to AAMI barrier classifications.

COMPETING INTERESTS

The authors declare that they have no competing interests.

AUTHORS' CONTRIBUTIONS

VP has contributed design and development of study and analysis of data. GT has carried out testing of samples. All authors read and approved the final manuscript.

REFERENCES

1. Association for the Advancement of Medical Instrumentation. (2003). *Liquid barrier performance and classification of protective apparel and drapes intended for use in health care facilities (Association for the Advancement of Medical Instrumentationthed.)*. Arlington, VA: ANSI/AAMI PB70.

2. Gawish, SM, Ramadan, AM, Matthews, SR, & Bourham, MA. (2008). Modification of PA 6,6 by atmospheric plasma and grafting hydroxy ethyl methacrylate to improve fabric properties. *Polymer Plastics Technology and Engineering, 47*(5), 473–478.

3. Huang, W, & Karen, L. (2007). Evaluating a one-bath process for imparting antibacterial and repellency to Nonwoven surgical gown fabrics. *Textile Research Journal, 70*(9), 774–782.

4. Huang, W, & Leonas, KK. (1999). One-bath application of repellent and antimicrobial finishes to nonwoven surgical gown fabrics. *Textile Chemist and Colorist, 31*(3), 11–16.

5. Jessie, H, Darlene, M, & Doris, H. (2007). Antibacterial and LaunderingProperties of AMS and PHMB as Finishing Agents on Fabric for Healthcare workers Uniforms. *Clothing & Textile Research Journal, 25*(3), 258–272.

6. Loveday, HP, Wilson, JA, Hoffman, PN, & Pratt, RJ. (2007). Public perception and the social and microbiological significance of uniforms in the prevention and control of healthcare-associated infections: an evidence review. *British Journal of Infection Control, 8*(4), 10–21.

7. Mews, P. (2009). Establish and maintain the sterile field Competency for safe patient careduring operative and invasive procedures. *Denver, 1*, 240–249.

8. Midha, VK, Vashisht, R, & Midha, V. (2014). Durability of fluoropolymer and antibacterial finishes on woven surgical gown fabrics. *Fashion and Textiles, 1*, 1–12.

9. Rajpreet, K, Gita, R, Mohamed, B, & Lee, B. (2004). Plasma and Antimicrobial treatment of Nonwoven fabric for surgical gowns. *Textile Research Journal, 74*(12), 1073–1079.

10. Rutala, WA, & Weber, DJ. (2001). A review of single use and reusable gowns and drapes in health care. *Infection Control and Hospital Epidemiology, 22*, 248–257.

11. Souma, D. (2012). Water and Stain Repellent Textile using Plasma Technology. *Journal of Applied Polymer Science, 1*(1), 11–24.

12. Tyner, DW. (2007). Evaluation of Repellent Finishes Applied By Atmospheric Plasma.*Journal of Plasma Science, 1*, 40–52.

13. Unsal, E, Dane, JH, & Schwartz, P. (2005). Effect of Liquid characteristics on the wetting, capillary, migration and retention properties of fibrous polymer networks. *Journal of Applied Polymer Science, 97*(1), 282–292.

14. Vaananen, R, Heikkila, P, Tuominen, M, Kuusipalo, J, & Harlin, A. (2010). Fast and efficient surface treatment for nonwoven materials by atmospheric pressure plasma.*AUTEX Research Journal, 10*(1), 8–13.

Index

A

ambulatory monitoring, 174
anisotropy, 92, 112, 114, 116
Antibacterial, 272, 280, 281, 286
artery, 91, 92, 93, 94, 95, 96, 98, 99, 101, 103, 104, 107, 108, 109, 112, 114, 115, 118, 121, 122, 123, 124, 176, 183
Automotive textiles, 150

B

bioethanol, 253, 254, 255, 256, 260, 262, 263, 266, 268
bioreactor material, 254, 256, 266
braiding, 91, 92, 93, 97, 104, 110, 123

C

cardiovascular disorders, 174, 193
compliance, 84, 92, 95, 96, 98, 99, 100, 101, 103, 106, 107, 110, 112, 114, 115, 118
conductive nanocomposites, 174

D

Denim fading, 38
dry electrodes, 174, 179, 180, 181, 187, 192, 193, 194

E

electrocardiography, 174, 190, 195
electrospinning, 91, 92, 93, 97, 107, 125, 127
Engraving, 38

F

Fashion, 37, 59, 61, 62, 129, 149, 197, 198, 199, 204, 206, 220, 225, 226, 271, 287
fermentation, 253, 254, 255, 256, 257, 258, 260, 261, 262, 263, 266, 267, 268
fiber Bragg grating sensors, 66, 84
fiber optic sensors, 2, 19, 23, 24, 25, 27, 28, 32, 33, 36, 65, 66, 74, 76, 80, 82, 87
Firefighters, 129, 130, 134, 135, 137, 138, 139, 141, 143, 144, 145, 146, 147

G

graft, 91, 92, 93, 97, 98, 99, 100, 101, 102, 103, 104, 107, 109, 112, 115, 117, 118, 119, 120, 121, 124, 125, 127, 275, 279

H

hetero-core fiber optics, 66
Hydrostatic, 272, 285

I

Impact penetration, 272, 283, 285
Incineration, 150
Inventory, 198, 204

K

knitting, 91, 92, 93, 97, 110

L

Landfill, 150
Laser, 37, 38, 39, 40, 41, 43, 44, 45, 46, 47, 48, 49, 51, 52, 57, 58, 59, 61, 62, 63, 84
Laser Fabric cutting, 38
low-melting-point polyester, 226, 227, 230

M

macrobending sensors, 66, 75
Magnetic Resonance Imaging, 66
Mass customization, 38, 46
Mobility, 130, 147
Moisture vapour, 272, 284, 285
MR-compatibility, 66, 69

N

nanotextiles, 174
non-linearity, 92, 95, 98, 112, 114
nonwoven geotextiles, 14, 226, 228, 229, 231, 232, 233, 234, 235, 236, 237, 239, 240, 241, 242, 243, 244, 246, 247

P

Protection Equipment, 130

R

reactor cost, 254
recycled Kevlar fibers, 226
Recycling, 150, 169, 171, 249, 250
respiratory monitoring, 25, 66, 69, 72, 74, 78, 79, 81, 86, 88
Retail, 198, 205, 217, 218
RFID, 50, 60, 61, 197, 198, 199, 200, 201, 202, 203, 204, 205, 206, 207, 208, 209, 210, 211, 212, 213, 214, 215, 216, 217, 218, 219, 220, 221, 222, 223, 224

S

Size Fit, 130
smart textiles, 65, 66, 67, 71, 72, 75, 79, 80, 81, 87, 174, 180, 187, 193
smartphone, 174, 185, 193
Supply chain management, 198
Sustainability, 149, 150, 154, 168, 171
Systems, 34, 60, 88, 149, 150, 152, 171, 173, 181, 191, 192, 195, 218, 219, 222, 224

T

tensile strength, 109, 116, 226, 227, 228, 229, 230, 231, 232, 233, 239, 243, 244, 246, 247, 256, 282, 283
textile bioreactor, 253, 254, 255, 256, 257, 258, 259, 260, 261, 262, 263, 264, 265, 266

V

vascular stent, 92
Viral penetration, 272, 276, 277, 280, 281

W

Waste, 150
Wearability Comfort, 130
wearable healthcare, 174
weaving, 91, 92, 93, 97, 110, 118, 178, 204